Push your Career Publish your Thesis

Science should be accessible to everybody. Share the knowledge, the ideas, and the passion about your research. Give your part of the infinite amount of scientific research possibilities a finite frame.

Publish your examination paper, diploma thesis, bachelor thesis, master thesis, dissertation, or habilitation treatises in form of a book.

A finite frame by infinite science.

Infinite Science Publishing

An Imprint of
Infinite Science GmbH
MFC 1 | Technikzentrum Lübeck
BioMedTec Wissenschaftscampus
Maria-Goeppert-Straße 1
23562 Lübeck
book@infinite-science.de
www.infinite-science.de

Editor

Thorsten M. Buzug
Institute of Medical Engineering
University of Lübeck
buzug@imt.uni-luebeck.de

Reihe: Medizinische Ingenieurwissenschaft und Biomedizintechnik

Diese Reihe umfasst Werke der Medizinischen Ingenieurwissenschaft und Biomedizintechnik, deren Themen strategisch unter den Zukunftstechnologien mit hohem Innovationspotenzial anzusiedeln sind. Als wesentliche Trends dieser Forschungsgebiete, sind die Schlüsselbereiche Computerisierung, Miniaturisierung und Molekularisierung zu nennen. Bei der Computerisierung sind dabei die inhaltlichen Schwerpunkte beispielsweise in der Bildgebung und Bildverarbeitung gegeben. Die Miniaturisierung spielt unter anderem bei intelligenten Implantaten, der minimalinvasiven Chirurgie aber auch bei der Entwicklung von neuen nanostrukturierten Materialien eine wichtige Rolle, und die Molekularisierung ist in der regenerativen Medizin aber auch im Rahmen der sogenannten molekularen Bildgebung ein entscheidender Aspekt. Forschungs- und Entwicklungspotenzial werden auch der Biophotonik und der minimal-invasiven Chirurgie unter Berücksichtigung der Robotik und Navigation zugeschrieben. Querschnittstechnologien wie die Mikrosystemtechnik, optische Technologien, Softwaresysteme und Wissenstechnologien sind dabei von hohem Interesse.

Henrik Rogge

In-Silico Analysis of Superparamagnetic Nanoparticles

Physics of the Contrast Agent
in Magnetic Particle Imaging

Medical Engineering Science and
Biomedical Engineering — Volume 12

Editor: Thorsten M. Buzug

Infinite Science
Publishing

© 2015 Infinite Science Publishing
der BioMedTec Wissenschaftsverlag Lübeck

Ein Imprint der Infinite Science GmbH,
MFC 1 | BioMedTec Wissenschaftscampus
Maria-Goeppert-Straße 1
23562 Lübeck

Cover Design, Illustration: Uli Schmidts, metonym
Copy Editing: University of Lübeck, Institute of Medical Engineering

Publisher: Infinite Science GmbH, Lübeck, www.infinite-science.de
Print: BoD, Norderstedt

ISBN Paperback:978-3-945954-13-3

Bibliografische Information der Deutschen Nationalbibliothek:
Die Deutsche Nationalbibliothek verzeichnet diese Publikation in der Deutschen Nationalbibliografie; detaillierte bibliografische Daten sind im Internet über http://dnb.d-nb.de abrufbar.

Bibliographic information published by the Deutsche Nationalbibliothek
The Deutsche Nationalbibliothek lists this publication in the Deutsche Nationalbibliografie; detailed bibliographic data are available in the internet at http://dnb.d-nb.de.

Zusammenfassung

Bildgebende Verfahren können als eine Erweiterung der menschlichen Sinne angesehen werden. Sie ermöglichen es Sachverhalte zu erforschen und zu beurteilen, welche ansonsten dem Menschen verborgen bleiben würden. Besonders in der Medizin sind bildgebende Diagnostikmethoden nicht mehr wegzudenken. Die schnelle Erfassung hoch komplexer Krankheitsbilder, welche auch nach gründlicher Anamnese und eingehender körperlicher Untersuchung nicht zugänglich wären, ist heutzutage eine tragende Säule in modernen Krankenhäusern.

Neben den gut etablierten medizinischen Diagnostikmethoden Ultraschall, Computertomographie und Magnetresonanztomographie gibt es seit kurzem Magnetic Particle Imaging (MPI). MPI ist in der Lage die Verteilung eines superparamagnetischen Kontrastmittels in einer hohen räumlichen und zeitlichen Auflösung abzubilden. Das Prinzip basiert auf dem nichtlinearen Magnetisierungsverhalten des Kontrastmittels, wobei dieser physikalische Effekt bis jetzt noch von keinem anderen bildgebenden Verfahren ausgenutzt wurde.

Diese Arbeit wurde am Institut für Medizintechnik in Lübeck angefertigt und setzt sich insbesondere mit den physikalischen Eigenschaften des MPI Kontrastmittels auseinander. Hierfür werden detailliert Informationen aus unterschiedlichen Bereichen zusammengefasst und aufbereitet. Des Weiteren wird ein Teilchenmodell basierend auf der Langevin Gleichung und der Fokker Planck Gleichung hergeleitet und physikalisch begründet. Hierbei stellt sich heraus, dass unter praktischem Gesichtspunkt die direkte numerische Lösung der Langevin Gleichung vielversprechender ist, als die Lösung der Fokker-Planck Gleichung.

Durch Lösen der stochastischen Differential Gleichungen wird dann das Frequenzverhalten des Kontrastmittels untersucht, wobei sich herausstellt, dass das Frequenzverhalten des Kontrastmittel nahezu linear ist und stark von der Anisotropie der Partikel abhängt.

Dieses stochastische Teilchenmodell ist in der Lage tiefe Einblicke in die Dynamik des Kontrastmittels zu geben. Jedoch erscheint es aufgrund seiner Komplexität und seiner Anforderung an Computer Rechenkapazität nicht als geeignet, um MPI Messdaten zu extrapolieren oder Kontrastmittel schnell zu charakterisieren. Deswegen beschäftigt sich diese Arbeit auch noch mit möglichen Teilchenmodellen, welche diese Aufgaben bewältigen können. Hierbei wird besonderes Augenmerk auf ein Teilchenmodell gelegt, welches auf der Lösung einer gewöhnlichen Differential Gleichung beruht.

Um die Parameter eines einfachen Teilchenmodells an MPI Messdaten anzugleichen, wird außerdem ein neuer genetischer Suchalgorithmus beschrieben. Dieser Suchalgorithmus ist in der Lage eine zweidimensionale, positive und nicht diskrete dabei aber beliebige Verteilung zu schätzen. Es zeigt sich, dass das auf gewöhnlichen Differential Gleichungen basierende Teilchenmodell und der genetische Suchalgorithmus fähig sind, Messdaten eines "Magnetic Particle Spectrometers" (Spektrometer für magnetische Partikel) abzubilden. Allerdings lässt sich leider die gefundene Lösung noch nicht auf Messdaten, welche mit einem zusätzlichen statischen Magnetfeld gemessen wurden, extrapolieren.

Da MPI auf der zentralen Annahme beruht, dass sämtliche Demagnetisierungsfelder und Teilchen-Teilchen Wechselwirkungen im Kontrastmittel zu vernachlässigen sind, wird in einem weiteren Kapitel die Gültigkeit dieser Annahme abgeschätzt. Dies wird mit Messergebnissen teilweise belegt.

Der Nutzen dieser Arbeit besteht vor allem in der detaillierten Herleitung des stochastischen Teilchenmodells und der vielversprechenden Kombination eines genetischen Suchalgorithmus mit einem einfachen Teilchenmodell, welches in Zukunft nützlich sein kann, um die MPI System Funktion zu generieren.

Das Verhältnis der Philosophie zur so genannten positiven Wissenschaft lässt sich auf die Formel bringen:
Philosophie stellt diejenigen Fragen, die nicht gestellt zu haben die Erfolgsbedingung des wissenschaftlichen Verfahrens war. Damit ist also behauptet, dass die Wissenschaft ihren Erfolg unter anderem dem Verzicht auf das Stellen gewisser Fragen verdankt.

(Carl Friedrich von Weizsäcker)

Contents

1 Motivation

Imaging technologies are extensions of the human senses. They visualize structures or processes by taking advantage of certain physical effects. For example, computer tomography displays the X-ray absorption properties of different tissues and medical ultrasonography measures the variable acoustical impedances tissues. However, there are often fixed limits of these imaging technologies, which are specifically caused by the underlying physical principle.

This emphasises the relevance of the recently proposed imaging technology Magnetic Particle Imaging (MPI), since it makes use of the non-linear magnetization response of iron oxide nanoparticles. This has not yet been used for imaging and therefore potentially extends our horizont.

This thesis deals with the physical properties of iron oxide nanoparticles with respect to MPI. In chapter two an introduction is given to the basic idea of MPI and in chapter three the physical properties of the associated contrast are summarized.

The fourth chapter presents a detailed derivation and validation of a particle model, which is based on a Langevin equation approach. In that chapter a numerical solver is introduced, which is able to handle the two diffusion processes, the Néel and the Brownian diffusion. The Néel diffusion describes the magnetic reversal due to the spins within the particle and the Brownian diffusion describes the rotational diffusion of the particle itself.

The fifth chapter deals with possible approximated particle models, which are important for MPI, since they promise to allow a faster acquisition of the MPI system function and a fast characterization of the performance of a specific contrast agent. Furthermore, a smooth two dimensional genetic fitting algorithm is described, which can be used to build up the connection between the measurement data and the approximated particle model.

In chapter six, a frequency sweep is simulated to gain insights into the dynamics of the MPI contrast agent and to investigate, whether there is an optimal frequency, which should be applied in MPI.

Since MPI basically assumes that particle-particle interactions of the nanoparticles in the contrast agent are negligible, a rough estimate is made in chapter seven up to which particle concentration this assumption holds.

In chapter eight some measurement results of a magnetic particle spectrometer are presented and the results, which have been obtained by fitting an approximated particle model to the measurement data. This work finally concludes with a summary and outlook.

2 Magnetic Particle Imaging

Magnetic particle imaging (MPI) is a new contrast agent based imaging technology, which takes advantage of the non-linear magnetization behavior of ferro- or ferrimagnetic nanoparticles. It is a very young research field, which has recently been invented by B. Gleich and J. Weizenecker in 2005 [2]. MPI promises to have a high sensitivity combined with a high spatial resolution [31]. In addition, MPI allows a fast image acquisition. In a first in vivo experiment, J. Weizenecker et al showed, that it is even possible to scan the cardiac cycle of a mouse [32], whereby the heart of a mouse beats at least 10 times per second. Since MPI intrinsically needs tracers it offers the possibility to gain information about physiological or pathological processes for example by functionalizing these particles with antibodies [71] or by visualizing the blood flow. On the other hand, it is only capable to uncover indirectly anatomical structures, which may be important landmarks for diagnosis.

Compared to other medical imaging technologies, such as computer tomography or positron emission tomography, the patient will not be exposed to x-rays or to radioactive radiations. Concerning MPI, one has to take care of two main toxic side effects. The patient-heating, induced due to the time varying magnetic field and toxic side effects, which may rise up due to the interaction of the contrast agent with the human body. Fortunately, both side effects have already been well studied, since the well-established magnetic resonance tomography uses time varying magnetic fields and the same contrast agent.

The research field of MPI may be separated into three main topics: MPI scanner development, image reconstruction techniques and research, which is concerned with the contrast agents. This diploma thesis focuses on the physical properties of the contrast agent and its dynamic.

2.1 Basic Idea of MPI

The MPI contrast agents are ferrofluids, which are stable suspensions of ferri- or ferromagnetic particles. The cores consist most often of Magnetite Fe_3O_4 and are surrounded by a shell, which prevents for agglomeration. The coating also ensures biocompatibility and it is often made of dextrans. The particle core diameter is about 10 nm and therefore one can expect, that it only consists of one single magnetic domain (3.2.4). This fact leads to a non-vanishing dipole moment $\boldsymbol{\mu}$ of each particle. If a homogeneous magnetic field $\boldsymbol{H}$ is applied to a ferrofluid, the magnetic moment of

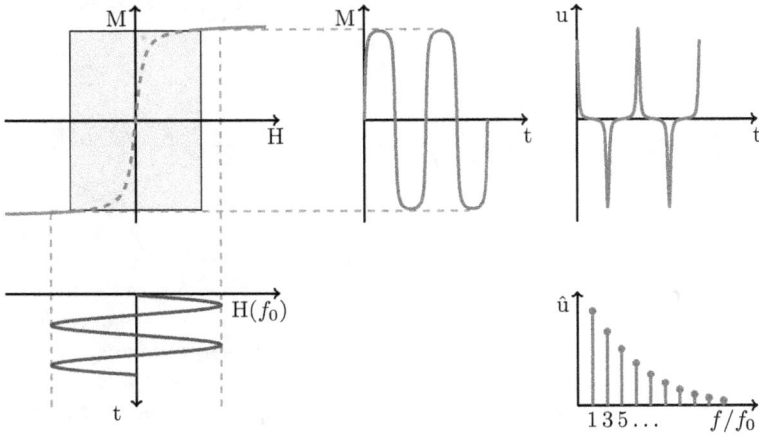

Fig. 2.1: The non-linear magnetization response M of a ferrofluid sample subjected to a time-varying magnetic field H with a fix base frequency f_0 will contain higher harmonics of f_0. Therefore the induced signal u will contain higher harmonics of f_0 and these higher harmonics can be detected within the spectrum $\hat{u}$.

the particles tries to align parallel to the magnetic field. This ordering process is disturbed by thermal agitation. Here the ratio $\rho = \mu_0|\boldsymbol{\mu}|\boldsymbol{H}/(k_\text{B}T)$ of the Zeeman energy $\mu_0|\boldsymbol{\mu}|\boldsymbol{H}$ to the thermal energy $k_\text{B}T$ is of importance. $\mu_0 = 4\pi \cdot 10^{-7}\,\text{Tm/A}$ is the vacuum permeability, $k_\text{B} \approx 1.38 \cdot 10^{-23}\,\text{J/K}$ is the Boltzmann constant and T is the temperature. If the ratio $\rho \gg 1$ is large, all magnetic moments are aligned with the magnetic field and the magnetization $\boldsymbol{M}$ of a ferrofluid probe is said to be saturated. The saturation magnetization $\boldsymbol{M}_\text{sat}$ can be calculated by

$$\boldsymbol{M}_\text{sat} = \frac{1}{V_\text{s}} \sum_{i=1}^{N} |\boldsymbol{\mu}|e_\text{H} \quad . \tag{2.1}$$

Here N is the number of particles in the sample volume V_s and $\boldsymbol{e}_\text{H}$ denotes the unit vector in the direction of the applied magnetic field. The magnetization between both saturated stats is non-linear. This is the dynamic region, which is marked yellow in figure (2.1) and (2.2). Its size can be characterized by the strength of the magnetic field H_dr, which is needed to saturate the contrast agent. So, the dynamic region spans over the interval $[-H_\text{dr}, H_\text{dr}]$. The magnetization behavior within the dynamic region can be calculated exactly, if the magnetic field is constant in time (3.4). But if the magnetic field varies in time, the magnetization exhibits rate-dependent hysteresis phenomena, because of the finite relaxation times of the particles (3.3).

The size of the dynamic region depends on the averaged magnetic moment of the particles. The magnetic moment of each particle is approxamtly given by

$$|\boldsymbol{\mu}| = V_\text{c}M_\text{s} \quad , \tag{2.2}$$

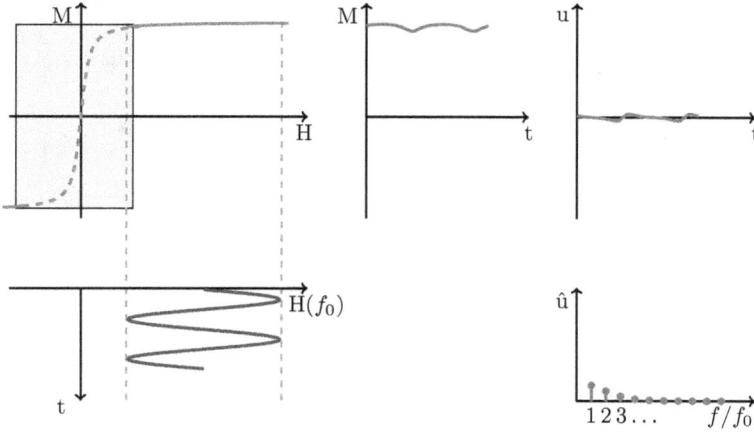

Fig. 2.2: The magnetization response of a ferrofluid sample vanishes if an additional, sufficiently strong offset magnetic field is applied. This in turn leads to a disappearance of higher harmonics in the spectrum of u.

whereas V_c is the core volume of the particle and M_s is the bulk saturation magnetization of the core material. Thus, the size of the dynamic region depends linear on the core volume of the ferrofluid-particles. H_{dr} is typical in the range of $H_{dr} \approx 10 - 50\,\text{mT}/\mu_0$.

If a time-dependent external magnetic field $\boldsymbol{H}_{ext}(\boldsymbol{r}, t)$ is applied to a ferrofluid sample, the magnetization of the sample will also depend on time. This magnetization signal can be recorded with an induction coil, whereas the induced voltage $u(t)$ is determined by the following integral [36]:

$$u(t) = -\mu_0 \int_\Omega \partial_t \boldsymbol{M}(\boldsymbol{H}_{ext}(\boldsymbol{r}, t), \boldsymbol{r}, t)\boldsymbol{p}(\boldsymbol{r})\mathrm{d}r^3 \tag{2.3}$$

$\boldsymbol{p}(\boldsymbol{r})$ is the sensitivity of the receiving coil, $\boldsymbol{H}_{ext}(\boldsymbol{r}, t)$ is the external applied magnetic field and Ω is the volume of interest. The sensitivity of a coil is defined by:

$$\boldsymbol{p}(\boldsymbol{r}) = \frac{\boldsymbol{H}_c(\boldsymbol{r})}{i_0} \tag{2.4}$$

Here $\boldsymbol{H}_c$ is the magnetic field, which would be produced by the coil at $\boldsymbol{r}$, if the coil would be driven by the current i_0. If the external magnetic field and the sensitivity of the receiving coil are homogenous and parallel over the sample volume V_s and the ferrofluid probe is homogeneous over V_s as well, the equation (2.3) simplifies to:

$$u(t) = -\mu_0 V_s p\, \partial_t \boldsymbol{M}(H_{ext}(t)) \tag{2.5}$$

MPI uses the non-linear magnetization behavior by applying a harmonic external magnetic field, called drive field H_D,

$$H_{ext}(t) = H_D(t) = A_D \sin(2\pi f_0 t) \tag{2.6}$$

to the ferrofluid probe(figure 2.1). If the amplitude A_D is sufficiently large to cover the dynamic region of the contrast agent, the relationship between the magnetic excitation field and the measured signal is non-linear. Because of this non-linear relationship, the Fourier series

$$u(t) = \sum_{n=-\infty}^{+\infty} \hat{u}_n e^{in\omega_0 t} \qquad (2.7)$$

contains higher harmonics of f_0. Here $\omega_0 = 2\pi f_0$ has been used and the Fourier coefficients can be calculated by

$$\hat{u}_n = \frac{\omega_0}{2\pi} \int_0^{2\pi/\omega_0} u(t) e^{-in\omega_0 t} \mathrm{d}t \quad . \qquad (2.8)$$

If an additional constant magnetic field $H_{\mathrm{const}} > H_{\mathrm{dr}} + A_D$ is applied to the system

$$H_{\mathrm{ext}} = H_{\mathrm{const}} + A_D \sin(2\pi f_0 t) \, , \qquad (2.9)$$

the magnetization dynamic vanishes, since the contrast agent is saturated. This in turn leads to the disappearance of the higher harmonics, which is visualized in figure (2.2). Thus, a ferrofluid sample can be detected, by detecting higher harmonics of the measured signal.

The spatial encoding is then achieved by using a gradient magnetic field $\boldsymbol{H}_S(\boldsymbol{r})$, which is superimposed by a harmonic drive field $\boldsymbol{H}_D(t)$. Only particles, that are in regions, where $|\boldsymbol{H}_S(\boldsymbol{r})| < |\boldsymbol{H}_D(t)| + H_{\mathrm{dr}}$ is valid contribute to higher harmonics. This area is called the field of view (FOV) and the measured spectrum is therefore linked to the spatial distribution of the contrast agents. $\boldsymbol{H}_S$ is often called a selection field, because it selects the regions, where the particles are not saturated and have a non-linear magnetization behavior. The easiest selection field is a magnetic field, whose field strength vanishes at a defined spatial coordinate. This leads to the concept of the field free point (FFP). Such a field can be generated in the center line of two Maxwell coils. However, there are even more advanced scanner designs, which make use of a field free line (FFL) [39, 74].

An image may be now generated by changing the selection field and therefore moving the FOV through space. Assuming a gradient strength of $G = 2\,\mathrm{T/m}$ and an excitation amplitude of $A_D = 10\,\mathrm{mT}$, one can roughly estimate the spatial resolution of a FFP scanner to be

$$\Delta x \approx \frac{2(H_{\mathrm{dr}} + A_D)}{G} = 2 - 6\,\mathrm{mm} \quad . \qquad (2.10)$$

This estimate shows, that the spatial resolution crucially depends on the size of the dynamic range, the strength of the gradient of the selection field and the amplitude of the drive field.

Fortunately, the measured spectrum exhibits a diverse structure, which strongly depends on the spatial distribution of the ferrofluid within the FOV. This additional information can be used to significantly increase the spatial resolution. Here a resolution below $\Delta x \leq 1\,\mathrm{mm}$ seems to be possible. Furthermore the additional information might be used in future to gain insight into different properties of the surrounding medium, for example the temperature or the viscosity [59, 58, 57].

If the ferrofluid is very low in concentration, particle-particle interactions are negligible and the magnetization can thus be written as:

$$M(H_{\text{ext}}(r,t),r,t) = c_{\text{N}}(r)\langle \mu(H_{\text{ext}}(r,t),t)\rangle \tag{2.11}$$

Here $c_{\text{N}}(r)$ is the particle concentration distribution, which for simplicity is assumed to be stationary in time. $\langle \mu(H_{\text{ext}}(r,t),t)\rangle$ is the mean magnetic moment, averaged over a large particle ensemble of non-interacting particles. It can be regarded as a single particle model.

With (2.11) and (2.3) one can reformulate (2.8) to

$$\hat{u}_n = \int_\Omega s_n(r)c(r)\mathrm{d}r^3 \quad , \tag{2.12}$$

whereas

$$s_n(r) = -\frac{\mu_0\omega_0}{2\pi}\int_0^{2\pi/\omega_0}\partial_t\langle \mu(H_{\text{ext}}(r,t),t)\rangle p(r)\mathrm{d}t \tag{2.13}$$

has been introduced. $s_n(r)$ is named the system function, since it contains all information about the used particles and magnetic field geometries. The continuous signal equation (2.12) can be discretized, by discretizing the space Ω into K equal-sized volumes ΔV.

$$\hat{u}_n \approx \Delta V \sum_{k=1}^K s_n(r_k)c(r_k) \tag{2.14}$$

If N harmonics are taken into account, this equation may then be expressed as a matrix-vector equation [36]

$$u = Sc, \tag{2.15}$$

where the vectors are given by

$$\begin{aligned} u &= (\hat{u}_n)_{n=1}^N \in \mathbb{C}^N \quad , \tag{2.16}\\ c &= (c(r_k))_{k=1}^K \in \mathbb{R}^K \tag{2.17} \end{aligned}$$

and the system matrix is given by

$$S = (\Delta V s_n(r_k))_{n=1\ldots N;\, k=1\ldots K} \in \mathbb{C}^{N\times K} \quad . \tag{2.18}$$

A more detailed distribution of the ferrofluid particles within the FOV may therefore be estimated by solving the inverse problem

$$c = S^+u \tag{2.19}$$

where S^+ is the pseudoinverse matrix of S. The solution of this inverse problem depends critically on the system matrix S. In general The system function can not be calculated analytically. Even approximate solutions with idealized magnetization functions or magnetic fields are up to now not possible. For example, the most simplified idealized magnetization function would be

$$\boldsymbol{M}(\boldsymbol{H}) = M_0 \begin{cases} \frac{\boldsymbol{H}}{|\boldsymbol{H}|} & \text{for } |\boldsymbol{H}| \neq 0 \\ 0 & \text{for } |\boldsymbol{H}| = 0 \end{cases} \quad . \tag{2.20}$$

and a good idealization of the magnetic fields is to consider only ideal homogenous fields or magnetic fields with a linear gradient. The system function can only be calculated in the case of an ideal one-dimensional MPI, whose magnetic field configuration is described by

$$\boldsymbol{H}_{\text{ext}} = \boldsymbol{H}_{\text{S}} - \boldsymbol{H}_{\text{D}} = \begin{pmatrix} G & 0 & 0 \\ 0 & G & 0 \\ 0 & 0 & 2G \end{pmatrix} \boldsymbol{r} - \begin{pmatrix} 0 \\ 0 \\ A\cos(\omega_0 t) \end{pmatrix} \tag{2.21}$$

and

$$\boldsymbol{p}(\boldsymbol{r}) = \boldsymbol{p}_0 \ . \tag{2.22}$$

Rahmer et al [30] proved, that the nth harmonic of the system function is a convolution

$$\begin{aligned} s_n(\boldsymbol{r}) &= -\frac{i\omega_0}{\pi} \boldsymbol{p}_0 \int_{-A}^{A} \frac{\partial \boldsymbol{M}(\boldsymbol{H}_{\text{S}}(\boldsymbol{r}) - \boldsymbol{H}_{\text{D}})}{\partial H_z} \mathrm{U}_{(n-1)}(H_{\text{D}}/A)\sqrt{1 - (H_{\text{D}}/A)^2}\mathrm{d}H_{\text{D}} \\ &= -\frac{i\omega_0}{\pi} \boldsymbol{p}_0 \frac{\partial \boldsymbol{M}(\boldsymbol{H}_{\text{S}}(\boldsymbol{r}))}{\partial H_z} * \left(\mathrm{U}_{(n-1)}(2Gz/A)\sqrt{1 - (2Gz/A)^2} \right) \end{aligned} \tag{2.23}$$

with the $(n-1)$th Chebyshev polynomial $\mathrm{U}_{(n-1)}$ of the second kind. However, this is no longer correct, if the finite relaxation times of the particles are included. Nevertheless, it shows that the system function is highly structured.

2.2 MPI Experiments

Until now, only a few MPI experiments have been conducted. All of them have to overcome quite challenging tasks, that are mainly caused by the coupling of the receiver and transmitter signal and the required bandwidth of the receiver signal [66].

Within each MPI experiment, the transmitter and the receiver coils are very close together, because both have to achieve high field strengths or sensitivities in the same region. Therefore, not only the time-varying magnetization of the contrast agent induces a signal in the receiver coil, but also the time-varying magnetic field produced by the transmitter coil. Unfortunately the signal of the particles is 10^6 to 10^9 times weaker and thus one has to separate both signals [36, 7].

Since MPI is based on the detection of higher harmonics of the base frequency, the signal receive chain has to handle a large bandwidth. For example, if one intends to detect 40 harmonics of the base frequency $f_0 = 25\,\mathrm{kHz}$, the bandwidth needs to range from $25\,\mathrm{kHz}$ to $1\,\mathrm{MHz}$.

Basically, the signal chain of each MPI device is identicall. The principle signal flow is presented in figure (2.3). It can be separated into a transmitter and a receiver part.

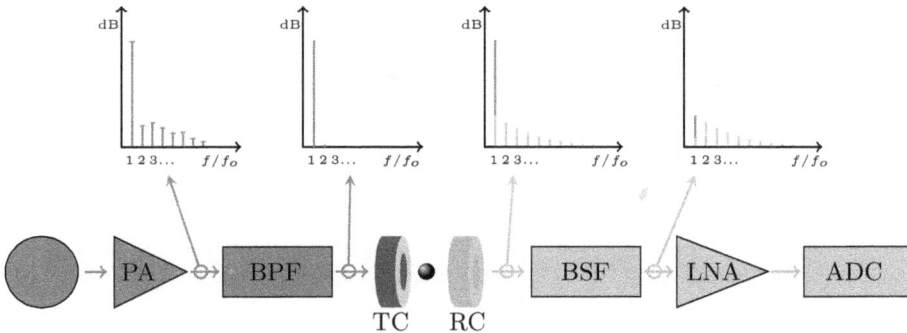

Fig. 2.3: Principle signal flow of an MPI device. A signal generator produces a signal, which is amplified by a power-amplifier (PA). The amplified signal does not only consists of the base frequency and therefore it has to be processed by a band-pass filter (BPF) before it is supplied to the transmitter coil (TC). The particle signal is picked up by a receiver coil (RC) and has to pass a band-stop filter (BSF), to get rid of the base frequency. The resulting signal is amplified by a low-noise amplifier (LNA) and digitized by an analog-to-digital converter (ADC).

The transmitter part starts with a signal generator, which in the majority of cases generates a sine-function at a fixed frequency f_0. Up to now the base frequency is most often to be around $f_0 \approx 25\,\mathrm{kHz}$ [32, 22]. However there are also MPI devices, which apply lower [23, 59] or higher base frequencies [55]. The signal generator is then followed by a power amplifier. Since there are no ideal signal generators or power amplifiers (PA), the amplified signal will not only consist of the fixed excitation frequency. All other frequencies have to be removed by a band-pass filter (BPF), before supplying the signal to the transmission coils (TC).

The receiver part starts with the pick-up coil. There are attempts to supplement additional induction coils, which are arranged in a way, so that they only see the magnetic field of the transmission coils, but not the magnetization signal of the particles. They are often called compensation coils and their signal might be subtracted of the signal from the ordinary pick-up coils [52]. Nevertheless it is inevitable that the signal of the transmission coils couples into the receiver coils and therefore the detected signal has to be processed by a band-stop filter (BSF), which extracts the high power transmitting frequencies. Afterwards, a low-noise amplifier (LNA) has to amplify the very weak signal, so that it can finally be digitized by an analog-to-digital converter (ADC). The LNA has to handle a large bandwidth, which is a challenging task [27].

2.2.1 Magnetic Particle Spectrometer

The magnetic particle spectrometer (MPS) is the simplest setup of a MPI device. It consists of one receiver coil and one pair of transmitter coils that apply a homogenous magnetic field to the ferrofluid sample. A MPS is used to study and characterize the contrast agents.

The MPS is actually not an imaging device, since it has no gradient field which could connect the particle magnetization response to the spatial distribution of the contrast agent. However, one can at least simulate an ideal one-dimensional MPI scanner by sweeping an additional static magnetic field over a certain range. A basic setup of a MPS is shown in figure (2.4).

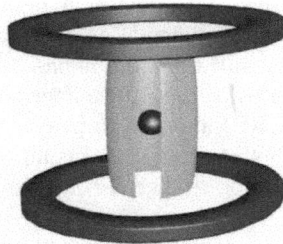

Fig. 2.4: Schematic setup of an MPS device. Two transmitter coils (blue) apply a homogenous magnetic field to a small ferrofluid probe. The magnetization response is detected by a solenoid receiver coil (orange).

2.2.2 3D MPI Scanner

The 3D MPI scanner proposed by B. Gleich and J. Weizenecker [2] is an FFP scanner, whose selection field is produced by two permanent magnets. The drive field is applied by three Helmholtz coil pairs that are mounted orthogonal to each other. The

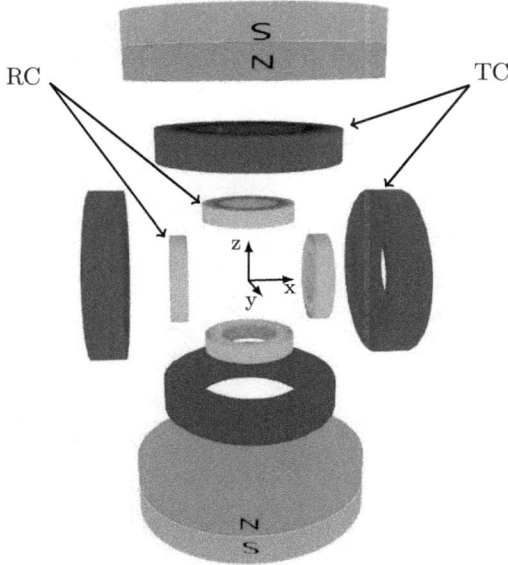

Fig. 2.5: Geometrical coil configuration of a 3D FFP scanner. The selection field is generated by two permanent magnets, which are aligned antiparallel. The transmitter coil pairs (TC) are Helmholtz coils, which are mounted orthogonal to each other. The signal is picked up by three receiver coil pairs (RC), which are also mounted orthogonal to each other. In order to reduce the graphical complexity, the receiver and transmitter coils in y direction have been omitted.

position of the FFP changes with time due to superposition of the static selection field with the time varying drive fields. The trajectory of the FFP depends on the applied transmission signals. All experiments published up to now use a Lissajous trajectory, which is generated by applying three sine-functions with different frequencies to the three transmitter coils TC_x, TC_y and TC_z. Any different trajectories may be obtained by using the appropriate drive field signals. But it turns out that the Lissajous trajectory is the optimal trajectory with respect to image resolution and image acquisition time [37]. However, this study is based on a static particle model, which excludes any rate-dependent hysteresis effects.

The three receiver coil pairs are aligned with the transmitter coils and record the particle signal generated within the FOV.

2.3 Acquisition of the System Function

The acquisition of the system function of a given MPI scanner and the corresponding contrast agent is a crucial point of MPI. On the one hand, the system function has to be quite exact, since the accuracy of the system function determines the results of the inverse problem (2.19). On the other hand, the acquisition of the system function has to be fast, since the system function cannot yet be transferred from scanner to scanner easily. Each time the magnetic coil setup or the contrast agent changes, the new system function has to be obtained. Even if the MPI scanner and the contrast agent remain unchanged and only the applied trajectory is changed, the system function changes. In future, it might be possible that different medical problems require lot of different contrast agents, for example labeled with different antigens, and different trajectories. Therefore, it is important to have a fast and exact acquisition method of the system function.

Measurement-Based Approach

One may obtain the system function by measuring the signal of a small delta sample, which is moved step-by-step from space point to space point [40]. If the sample volume ΔV and the concentration c_0 of the ferrofluid is known, the measured signal of the delta sample at the space point $\boldsymbol{r}_k$ can be calculated by

$$\hat{u}_n(\boldsymbol{r}_k) = \int_\Omega \delta(\boldsymbol{r}_k) s_n(\boldsymbol{r}_k) c(\boldsymbol{r}_k) \mathrm{d}r^3 \approx s_n(\boldsymbol{r}_k) c_0 \Delta V \qquad (2.24)$$

and therefore the system matrix is given by

$$s_n(\boldsymbol{r}_k) = \hat{u}_n(\boldsymbol{r}_k)/c_0 \Delta V \quad . \qquad (2.25)$$

This measurement technique automatically includes the complex particle dynamic and all different system parameter, which is a big advantage. But it also has an important drawback. The measured system function contains noise and the signal-to-noise ratio (SNR) scales proportionally to the ferrofluid concentration and the square root of the measurement time T_{meas}.

$$\mathrm{SNR} \propto c_0 \sqrt{T_{\mathrm{meas}}} \qquad (2.26)$$

Consequently each measurement point requires a minimum measurement time, which in total results in a very slow acquisition time of the whole system function. For example, even if restricted to a low spatial resolution of $20 \times 20 \times 20$ grid points and a quite short measurement time of $T_{\mathrm{meas}} = 4\,\mathrm{s}$, the acquisition of the system function takes about 9 hours. Therefore, it is not reasonable to gain a higher spatial resolution with this method.

Furthermore, due to the SNR relation, one would like to measure a ferrofluid sample with a high particle concentration. But of course, if the concentration of the contrast agent increases particle-particle interactions and demagnetization effects start to dominate. Related to this problem, some basic estimates are made in chapter (7).

Simulation-Based Approach

The simulation-based generation of the system function basically splits into two main tasks. On the one hand the simulation of magnetic fields and on the other hand the simulation of the magnetization response of the particle.

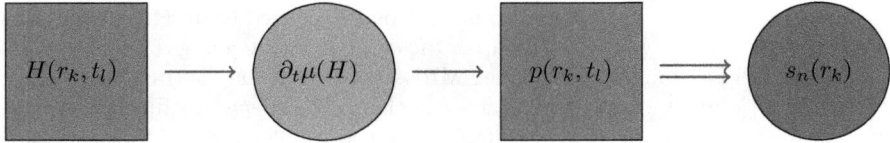

Fig. 2.6: Each MPI Simulation basically consists of three steps. First one has to simulate the magnetic fields, produced by the transmitter coils. Using these result it is possible to simulate the particle's magnetization response, which is finally picked up by the receiver coils, which are specified by their sensitivity field.

According to the Bio-Savart Law the magnetic field at space point r, generated by a current density $j(r', t)$ at space point r', can be calculated by

$$\boldsymbol{H}(\boldsymbol{r}, t) = \frac{1}{4\pi} \int \boldsymbol{j}(\boldsymbol{r}', t) \times \frac{\boldsymbol{r} - \boldsymbol{r}'}{|\boldsymbol{r} - \boldsymbol{r}'|^3} \mathrm{d}^3 r' \quad . \tag{2.27}$$

Furthermore, one may separate time and space by using the definition of the coil sensitivity.

$$\boldsymbol{H}(\boldsymbol{r}, t) = I(t)\boldsymbol{p}(\boldsymbol{r}) \tag{2.28}$$

However, since each MPI Scanner is a composition of several coils, this separation is no longer valid. Here, one has to include the coupling of the coils. The time-varying magnetic flux produced by a transmitter coil will of course also induce a voltage in all other coils. This in turn leads to a different simulated magnetic field as expected, if one would make use of (2.28).

Nevertheless, this problem is solvable [36], whereas it is still an unsolved question, how one should simulate the dynamic magnetization response of the contrast agent. Here it seems to be a good approach to simulate the interaction of the particles with the magnetic field and the thermal agitation according to a Langevin Equation Approach. This is presented in detail in (4).

However, even if there would be an implemented simulation method, which is capable to describe all present physical effects of an MPI Scanner in combination with its contrast agent, it seems to be hardly possible that a pure simulation-based acquisition of the system function leads to promising results. This is mainly due to the multiple properties of the contrast agents, which prevent a consistent characterization.

But of course, there is need for a good simulation model of a MPI device, since it will give a better physical insight into this new imaging technology. Therefore, one can expect that a verified simulation model will in future have large influence on the scanner and particle design.

Combined Method

Since both, the measurement- and simulation-based approach, have some advantages and disadvantages, a combined method might turn-out to be more promising.

The magnetic field simulation may be validated by measurements with a Hall probe or one may even think of measuring the whole magnetic field, since this has to be done only once for each scanner with a determined trajectory.

To include the particle signal, one may think of measuring the contrast agent in an MPS and then extrapolate the measurement data to any arbitrary scanner. Therefore, one needs a particle model, which is capable to handle the non-linear magnetization behavior combined with rate-dependent hysteresis effects.

The particle model derived by the Langevin equation approach should not be used, since it also contains noise and thus there is need for a new and easy to handle particle model. Such a particle model will contain parameters, which for example measure the magnetic moment or the relaxation time. These parameters may be estimated by relating them to MPS measurement data. Afterwards these parameterized particle model can be extrapolated to different scanner geometries. This approach promises to be fast and to generate a system function, which contains no noise. In (5) some particle models are investigated and a promising fitting algorithm is proposed.

3 Properties of the contrast agent

Magnetic Particle Imaging intrinsically needs a contrast agent. These contrast agents are ferrofluids, which are stable suspensions of nanometric-sized magnetic particles. Ferrofluids are in extensive use in technical application such as rotating shaft seals, viscous dampers and of course in medical applications [24].

One of the earliest investigators of ferrofluids was S. Papell, who worked in the beginning of the 1960's for the NASA. He pursued the plan to control rocket engine propellants viscosity in space, by adding a ferrofluid under the influence of a magnetic field [53].

The physical theory of ferrofluids and magnetic fluids was strongly influenced by E. Rosensweig [63].

3.1 Stability Criteria of Ferrofluids

Most of the ferrofluids used in medicine consist of Magnetite particles ($Fe_2^{3+}Fe^{2+}O_4$) with a core diameter d_c of about 10 nm. The particles are that small that thermal agitation prevents sedimentation in gravitational fields. In general, the stability of the colloidal suspension is very important for MPI, because this ensures the time independent quality of a ferrofluid sample [63].

A magnetic gradient field would also promote aggregation of ferrofluid particles and hence stability requires the dominance of the thermal energy over the magnetic energy:

$$\frac{\text{Thermal Energy}}{\text{Magnetic Energy}} = \frac{k_B T}{\mu_0 H M_s V_c} \geq 1 \tag{3.1}$$

The bulk saturation magnetization of magnetite is about $M_s = 494000\,\text{A/m} \approx 0.6\,\text{T}/\mu_0$ [73, 11] and additional reasonable values in the case of MPI would be $T = 300\,\text{K}$ and $H = 20\,\text{mT}/\mu_0$. One is therefore able to estimate the largest possible particle diameter

$$d_c \leq \left(\frac{6k_B T}{\mu_0 H M_s \pi} \right)^{\frac{1}{3}} \approx 10\,\text{nm} \quad . \tag{3.2}$$

However, sedimentation is not the only process, which endangers stability of a ferrofluid. The constant dipole moments of the particles provide the opportunity of

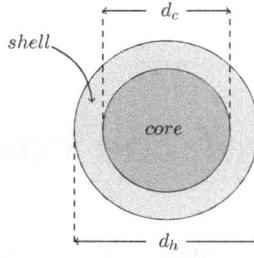

Fig. 3.1: The simplest model of a ferrofluid particle considers the particle to be spherical and that each particle core is surrounded by its own shell.

agglomeration. The energy between two dipoles is [13]

$$E_{\mathrm{dd}} = \frac{\mu_0}{4\pi} \left[\frac{\boldsymbol{\mu}_1 \boldsymbol{\mu}_2}{r^3} - \frac{3}{r^5} (\boldsymbol{\mu}_1 \mathbf{r})(\boldsymbol{\mu}_2 \mathbf{r}) \right] \quad . \tag{3.3}$$

The arrangement in particle chains is energetically favored. In the chain-like configuration the dipole-dipole interactions energy reduces with $\boldsymbol{\mu}_1 \boldsymbol{\mu}_2 = \boldsymbol{\mu}^2$ and $\boldsymbol{\mu}_1 \mathbf{r} \boldsymbol{\mu}_2 \mathbf{r} = \boldsymbol{\mu}^2 r^2$ to

$$E_{\mathrm{d||d}} = -\frac{\pi \mu_0 M_{\mathrm{s}}^2 d_{\mathrm{c}}^6}{72 r^3} \quad . \tag{3.4}$$

Each particle is coated with a shell and therefore the minimum distance is limited to the diameter of the whole particle. This diameter is often denoted as the effective hydrodynamic diameter d_{h}. Again, the relation to the thermal energy is of importance.

$$\frac{\text{Thermal Energy}}{E_{\mathrm{d||d}}} \frac{72 d_{\mathrm{h}}^3 k_{\mathrm{B}} T}{\pi \mu_0 M_{\mathrm{s}}^2 d_{\mathrm{c}}^6} \geq 1 \tag{3.5}$$

This leads to a quadratic equation

$$d_{\mathrm{c}}^2 - (d_{\mathrm{c}} + a_{\mathrm{shell}}) \sqrt[3]{\frac{k_{\mathrm{B}} T 72}{\pi \mu_0 M_{\mathrm{s}}^2}} = 0 \quad . \tag{3.6}$$

Clearly, a very thick shell prevents the particles from sticking together. A $a_{\mathrm{shell}} = d_{\mathrm{c}} - d_{\mathrm{h}} = 20\,\mathrm{nm}$ thick shell leads to the following estimate:

$$d_{\mathrm{c}} \leq 16\,\mathrm{nm} \tag{3.7}$$

These estimates show that it is difficult to produce stable ferrofluids and how important the surfactant is. A stable ferrofluid will be able to resolve itself after agglomeration or sedimentation, if for example external fields are switched of. But these considerations give a hint that the majority of particles is in the range of $d_{\mathrm{c}} \approx 10\,\mathrm{nm}$.

Additional, van der Waals like forces are of great importance. These are attractive and very short-range forces ($\propto r^{-6}$), but since it is very likely that particle collisions happen, due to the high particle concentration ($\approx 10^{14}\,1/\mu l$) of ordinary water based ferrofluids, they have to be included. Here, the shell prevents the particles from sticking together due to steric repulsion and Coulomb interactions. Especially, Coulomb forces depend strongly on the pH-value of the solvent.

3.2 Micromagnetic Properties of Nanoparticles

The magnetism of ferri- or ferromagnetic nanoparticles can be understood within the theory of micromagnetism. This theory is a combination of quantum mechanics and the Maxwellian theory of electromagnetic fields.

Quantum mechanical effects are important in the theory of magnetism, because exchange energy and crystal anisotropy do not have a classical analogon. In addition, the dimension of a nanoparticle is comparable to typical length scales. A pure quantum mechanical model however would not be solvable, since the nanoparticles contain up to 10^5 atoms. The quantum mechanical phenomena are thus embedded in a classical continuum theory [41]. The theory of micromagnetism has been developed since the 1930s by famous physicist like Brown, Heisenberg, Dirac, Bloch, Néel, Landau and Lifshitz.

3.2.1 Gibbs Free Energy Density

The starting point of micromagnetic calculations is the Gibbs free energy density [41]

$$\Phi = U - TS - \sigma\epsilon - \boldsymbol{M}(\boldsymbol{r}) \cdot \boldsymbol{H}_{\text{ext}} \quad . \tag{3.8}$$

Here U is the internal energy, T the temperature, S the entropy, σ the stress tensor and ϵ the strain tensor. $\boldsymbol{M}(\boldsymbol{r})$ stands for the space dependent magnetization and $\boldsymbol{H}_{\text{ext}}$ for the external applied magnetic field. Here the internal energy U includes the exchange energy and the magnetostatic interaction. The stress and strain tensors can be omitted, since the typical field strengths in MPI are far too low and the dimensions of the nanoparticles too small, so that magnetoelastic effects are not of importance.

In thermodynamic equilibrium the Gibbs free energy is at a minimum and thus the searched configuration of $\boldsymbol{M}(\boldsymbol{r})$ reduces Φ to a minimum.

3.2.2 Exchange Energy

The exchange energy is necessary to understand phenomena like the ordering structure of ferri- or ferromagnets. The classical dipole-dipole interaction energy is far too small to obtain this ordered alignment of magnetic moments and the thermal agitation would destroy these structures even at a few Kelvin [69].

The exchange interaction arises from the Pauli exclusion principle, which states, that it is not allowed, that two indistinguishable fermions of a system are in the same state. So the system wave function Ψ has to be antisymmetric if two fermions are permuted.

$$\Psi(r_1, s_1, r_2, s_2) = -\Psi(r_2, s_2, r_1, s_1) \tag{3.9}$$

Often the Coulomb interaction of an antisymmetric space wave function is smaller compared to the symmetric one. An antisymmetric space wave function leads due to the Pauli principle to a symmetric spin wave function. So the exchange interaction may be understood as an interaction of an overlap of wave functions, indistinguishability of particles and their Coulomb interaction.

An important model of exchange interaction of localized spins is the Heisenberg model. Its Hamiltonian is defined by

$$\widehat{\mathbf{H}} = -2 \sum_{(i,j)} J_{ij} \widehat{S}_i \widehat{S}_j \quad , \tag{3.10}$$

whereby J_{ij} is the exchange integral between the i-th and j-th spin. The spins alignment is parallel if $J > 0$ and anti-parallel if $J < 0$. Within the theory of micromagnetism the exchange interaction is introduced by the following term

$$\frac{E_{\text{ex}}}{V} = A(\nabla \mathbf{m})^2 \quad . \tag{3.11}$$

Where $\mathbf{m} = \frac{M}{M_s}$ is the reduced magnetization vector and A the exchange stiffness constant. In the case of a cubic lattice, A is given by

$$A = \frac{JS^2}{a} \quad . \tag{3.12}$$

The exchange constant J depends on the considered material, likewise the unit length a of the crystal lattice, whereas S is the magnetic moment of of an electron spin. The term $\nabla \mathbf{m}$ surpresses inhomogeneities of the spin arrangement and therefore supports a parallel spin alignment.

The Heisenberg Model is appropriate to understand basic exchange phenomena, but it is not capable to describe the magnetism of non-localized conduction electrons or some mechanisms in ferrimagnetic materials. For example, the indirect coupling of the two Fe^{3+} atoms through an oxygen atom in magnetite is described by the theory of superexchange.

3.2.3 Anisotropy

The magnetization behavior of a particle depends on its orientation to the external magnetic field. Nontheless the value of the saturation magnetization remains unchanged, since the number of magnetic moments is constant. This anisotropic behavior exists mainly because of the crystal anisotropy and the shape anisotropy of the particle. Furthermore, it is always invariant with respect to a 180° turn.

The crystal anisotropy arises due to the interaction between the magnetic moments of the spins and because of the orbital motions of the electrons. These L-S coupling displaces the orbits, which results in a favored pointing direction of the spins. The preferred direction is named "easy direction" or "easy axis".

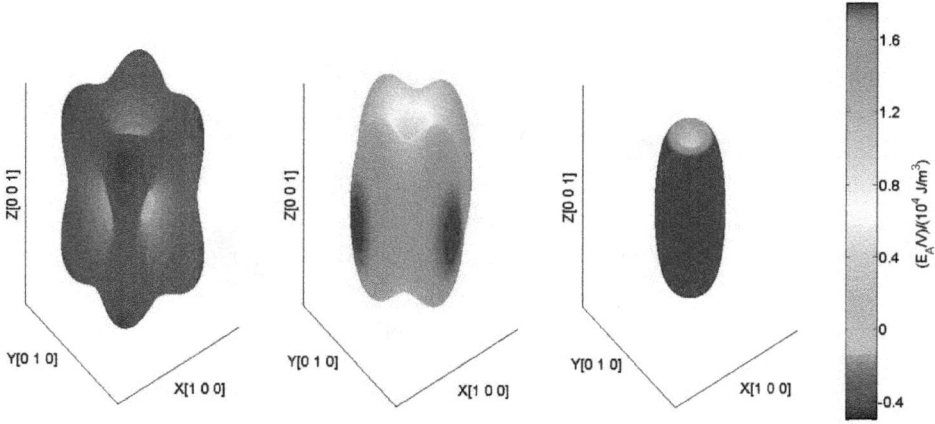

Fig. 3.2: This figure shows the anisotropy energy density E_A/V. The left figure visualizes the cubic crystal anisotropy of magnetite with $K_1 = -1.36 \cdot 10^4 \, \text{J/m}^3$ and $K_2 = -0.44 \cdot 10^4 \, \text{J/m}^3$. The right figure shows the energy surface of an uniaxial shape anisotropy, calculated with $K_S = 1.8 \cdot 10^4 \, \text{J/m}^3$. The figure in between visualizes the superposition of the shape and the crystal anisotropy energy. Here the easy axis coincides with the z axis of the crystal.

Because of the symmetry of the problem, it is common use to express the reduced magnetization vector in spherical coordinates [70]:

$$\mathbf{m}_x = \sin(\theta)\sin(\phi)\mathbf{e}_x \tag{3.13}$$
$$\mathbf{m}_y = \sin(\theta)\cos(\phi)\mathbf{e}_y \tag{3.14}$$
$$\mathbf{m}_z = \cos(\theta)\mathbf{e}_z \tag{3.15}$$

If the anisotropy energy depends only on one single axis (for simplicity the z-axis), it is possible to calculate the energy with respect to a single angle:

$$E_A = V_c \sum_{i=1}^{\infty} K_i \sin^{2i}(\theta) \tag{3.16}$$

This uniaxial anisotropy is often terminated after the first term, so that only one anisotropy constant K is involved

$$E_A = V_c K_1 \sin^2(\theta) \quad . \tag{3.17}$$

Besides the uniaxial anisotropy, there are also more complex anisotropy types. For example, magnetite has a cubic crystal anisotropy, which can be calculated with the following equation [11]

$$E_A = V_c \left(K_0 + K_1((m_x m_y)^2 + (m_y m_z)^2 + (m_x m_z)^2) + K_2(m_x m_y m_z)^2 + .. \right) \quad , \tag{3.18}$$

whereby the two anisotropy constants are given by $K_1 = -1.36 \cdot 10^4 \, \mathrm{J/m^3}$ and $K_2 = -0.44 \cdot 10^4 \, \mathrm{J/m^3}$ [45].

The shape anisotropy is induced by free poles on the surface of a particle. These poles induce a demagnetization field $\mathbf{H_d}$, which is proportional to the intensity of the magnetization $\boldsymbol{M}$ of the particle [10]

$$\mathbf{H_d} = \mathcal{N}\mathbf{M} \quad . \tag{3.19}$$

In the case of a single domain particle the intensity of the magnetization $\boldsymbol{M}$ is approximately given by the bulk saturation magnetization M_s of its core material.

Here, $\mathcal{N}$ is the demagnetization factor. In general, this factor is a tensor, which depends on the shape and the applied direction of the external magnetic field. If the particle is of ellipsoidal shape, the demagnetization field is homogeneous throughout the particle and it is possible to calculate the demagnetization factors analytically. For example, the demagnetization factor along the symmetry axis of an ellipsoid with a rotational cross section is given by [25]

$$\mathcal{N}_c = \frac{1}{\alpha^2 - 1} \left[\frac{\alpha}{2(\alpha^2 - 1)^{\frac{1}{2}}} \ln\left(\frac{\alpha + (\alpha^2 - 1)^{\frac{1}{2}}}{\alpha - (\alpha^2 - 1)^{\frac{1}{2}}} \right) - 1 \right] \quad . \tag{3.20}$$

Where the aspect ratio α of a rotational symmetric ellipsoid is defined as the ratio between the length of the symmetry axis c and the length of one of the other principle axis, see figure (3.3). Furthermore, the demagnetization factors of the principle axis of the ellipsoid have to fulfill the following condition

$$\mathcal{N}_a + \mathcal{N}_b + \mathcal{N}_c = 1 \quad . \tag{3.21}$$

The magnetostatic energy of a particle can be calculated by

$$E_{\mathrm{ms}} = -\frac{1}{2}\mu_0 \int_{V_c} \mathrm{d}V \mathbf{H_d}\mathbf{M} \quad . \tag{3.22}$$

This integral can be solved in the case of a rotational ellipsoid and the solution is given by

$$E_{\mathrm{ms}} = -\frac{1}{2}\mu_0 \mathcal{N}_c M_s^2 V_c - \frac{1}{4}\mu_0 M_s^2 V_c (1 - 3\mathcal{N}_c) \sin^2(\theta) \quad . \tag{3.23}$$

Note that only the demagnetization factor $\mathcal{N}_c$ occurs.

Both the crystal and the shape anisotropy contribute to the anisotropic behavior of the particles. The shape anisotropy can be compared to the crystal anisotropy, by calculating the shape factor $\frac{1}{4}\mu_0 M_s^2(1-3\mathcal{N}_c)$. Assuming an aspect ratio of $\alpha = 1.4$, one comes to the conclusion that in the case of magnetite the shape anisotropy dominates the crystal anisotropy.

$$\frac{1}{4}\mu_0 M_s^2 (1 - 3\mathcal{N}_c) \approx 1.8 \cdot 10^4 \frac{\mathrm{J}}{\mathrm{m^3}} > 1.36 \cdot 10^4 \frac{\mathrm{J}}{\mathrm{m^3}} = |K_1| \tag{3.24}$$

As presented in figure (3.2), the energy surface of the anisotropy can be very complex. Thus, it is common use to consider only particles with a uniaxial anisotropy and to introduce an effective anisotropy constant K_{eff}. K_{eff} for magnetite is usually taken to be between $10^4 \, \mathrm{J/m^3}$ and $10^5 \, \mathrm{J/m^3}$ [11].

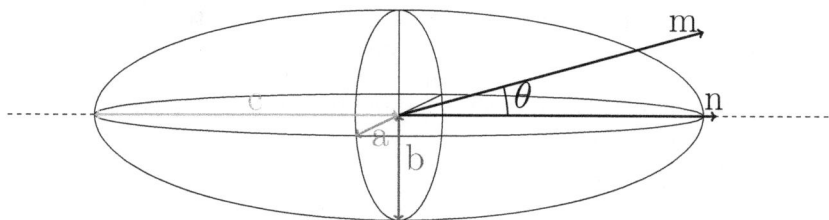

Fig. 3.3: Ellipsoid with a rotational cross section, here the effects of the shape anisotropy can be solely described by the angle θ between the easy axis $\boldsymbol{n}$ and the reduced magnetic moment $\boldsymbol{m}$. The aspect ratio is given by $\alpha = c/a = c/b$.

3.2.4 Magnetic Domains

Ferromagnetic and ferrimagnetic materials consist of magnetic domains, if their temperature is below the Curie temperature. Within these domains the spin alignment is homogenous, but if the whole specimen is not magnetized, the orientation of domains points in different direction and therefore the net magnetization is zero.

This phenomenon can be understood as a competition of the magnetostatic energy with the crystal anisotropy energy and the exchange energy. Many domains would decrease the magnetostatic energy, whereas the anisotropy energy and exchange energy would increase, and vice versa. The configuration of domains, which minimizes the Gibbs free energy density (3.8) wins. The anisotropy energy and exchange energy increase with the number of domains, because of the areas between the domains. These areas are called domain walls. Here the magnetization direction turns from one domain to the spin alignment of the next neighbor domain. The wall dimension is again determined by a competition between the anisotropy energy and the exchange energy. The exchange energy would prefer a thick wall, because this means that there is only a slight change in the arrangement of neighboring spins. But a thick domain wall contains many spins, which are not allowed to point in the easy directions of the crystal, so this would increase the anisotropy energy.

Considering a spherical particle, it is possible to calculate the critical diameter D_{cr} up to which it is barely energetically efficient to consist of a single domain. This critical diameter is solely characterized by material constants [41]

$$D_{\mathrm{cr}} = \frac{72\sqrt{AK}}{\mu_0 M_{\mathrm{s}}^2} \quad . \tag{3.25}$$

In literature one finds different values of the critical diameter of magnetite. These range from $12.4\,\mathrm{nm}$ [41], $70\,\mathrm{nm}$ [45] up to $30 - 100\,\mathrm{nm}$ [11].

Magnetic domains are strongly linked with the magnetic moment of each particle. Very small particles ($d_{\mathrm{c}} < 3\,\mathrm{nm}$) will not exhibit ferromagnetic of ferrimagnetic properties, since there are simply not enough atoms [12]. Intermediate particles have a smaller magnetic moment then predicted by $|\boldsymbol{\mu}| = M_s V_c$, which is due to surface effects. Very large single domain particles with a diameter just below the critical diameter may even have no magnetic moment. Because besides the single and the two domain state, there might be another state, called the vortex state. Here the spin alignment is no more parallel, but the spins are arranged in vortices [25]. Thus, the formula $|\boldsymbol{\mu}| = M_s V_c$ can only be regarded as an approximation.

3.3 Relaxation Processes in ferrofluids

There are two types of relaxation processes present in a ferrofluid. First the particle can rotate as a rigid body. This is named the Brownian rotation. And the second process is called Néel Rotation. It describes the degree of freedom of the magnetic moments to rotate with respect to the particle itself.

3.3.1 Brownian Rotation

The characteristic diffusion time of the Brownian motion is given by [75]

$$\tau_B = \frac{\xi}{\beta} = \frac{3\eta V_h}{k_B T} \quad . \tag{3.26}$$

Here η is the viscosity of the surrounding medium and V_h stands for the hydrodynamic Volume, which is assumed to be the whole particle including the shell.

The Brownian diffusion process does not depend on the inertia of the particle. This has been proven by Einstein [75]. He considered a particle, whose rotation is determined by the initial angular velocity ω_0 and the frictional term $\xi\omega$. Starting with

$$\dot{\theta}|_{t=t_0} = \omega_0 \tag{3.27}$$
$$I\dot{\omega} = -\xi\omega \tag{3.28}$$

it follows, that the time evolution of the angular velocity is given by

$$\omega(t) = \omega_0 \exp(-t\xi/I) \quad . \tag{3.29}$$

The inertia may be therefore neglected, if $\frac{I}{\xi}$ is small related to a characteristic measurement time. Neglecting the weight of the shell this quotient can be calculated and is given by

$$\frac{I}{\xi} = \frac{d_c^5 \rho_{Fe_3O_4}}{60\eta d_h^3} \approx 2 \cdot 10^{-10} \text{s} \ll 4 \cdot 10^{-5} \text{s} \quad . \tag{3.30}$$

With $d_c = 30\,\text{nm}$, $d_h = 35\,\text{nm}$ and $\rho_{Fe_3O_4} = 5200\,\text{kg/m}^3$. Compared to the typical MPI base frequency $1/25\,\text{kHz} = 4 \cdot 10^{-5}\,\text{s}$ the effect of inertia is negligible.

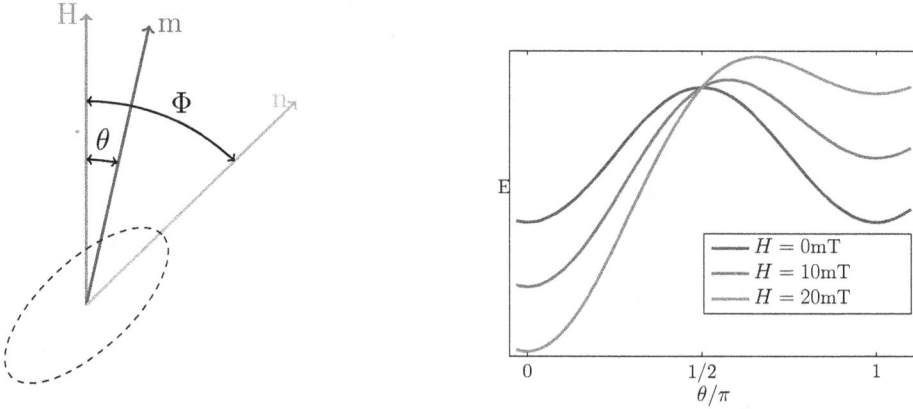

Fig. 3.4: The Stoner Wohlfarth particle model describes the magnetization behavior of a rigid particle. Φ is the angle between the easy axis n and the applied magnetic field H and θ is the angle beteween the magnetic field and the magnetic moment m of the particle. The right figure visualizes the energy surface, if the easy axis is aligned with the magnetic field.

3.3.2 Néel Rotation

The internal Néel diffusion can be illustrated with the Stoner Wohlfarth particle model and the Arrhenius law. The Stoner Wohlfarth particle model is simple and instructive and helps to understand the magnetization behavior of rigid particles. The anisotropy is assumed to be uniaxial and furthermore the spins within the particle are assumed to be parallel throughout the rotation, which leads to a vanishing exchange contribution ($\nabla m = 0$)(3.11). Therefore, the energy of such a particle is correctly described by the anisotropy and the Zeeman energy [1]

$$E = \underbrace{K_{\text{eff}}V \sin^2(\phi - \theta)}_{\text{Anisotropy Energy}} - \underbrace{\mu_0 M_{\text{s}}V H \cos(\phi)}_{\text{Zeeman Energy}} \quad . \tag{3.31}$$

θ is the angle of the magnetization vector to the magnetic field axis and ϕ is the angle between the magnetic field H and the easy axis. If there is no magnetic field, the two equilibrium states are separated by an energy barrier $\Delta E = K_{\text{eff}}V$, see figure (3.4). According to the Arrhenius law, Néel proposed to use the following equation to calculate the relaxation time of such a particle [75]

$$\tau_{\text{N}} = \tau_0 e^{KV_{\text{eff}}/k_{\text{B}}T} \quad , \tag{3.32}$$

where τ_0 is of the order of the gyromagnetic rotation frequency (10^{-11}s - 10^{-9}s).

It is noteworthy, that even a slight polydispersity of the particles leads, due to the exponential dependence, to a large variation of the diffusion time τ_{N}.

The assumption of coherent rotation is no more valid for large particles, because of a smaller exchange energy contribution. Larger particles may exhibit different magnetic reversal modes, whereas the spin alignment is no more parallel during the rotation. This in turn leads to a smaller magnetostatic energy and therefore to a smaller effective anisotropy [4]. One may therefore think of simulating large particles with a coherent rotation but with a smaller anisotropy ($K_{eff} < 10^4 \, J/m^3$).

The term "Superparamagnetism" is usually defined by the relationship of the Néel Diffusion time to a characteristic measurement time τ_m of an experiment [25]. If $\tau_m < \tau_N$, the magnetization of the particle is measured to be zero, because of the thermal fluctuations. Whereas in the case of $\tau_m > \tau_N$ the particle is said to be "blocked" and the magnetization does not level to zero, since the thermal agitation is not strong enough to overcome the energy barrier $\Delta E = K_{eff} V$.

According to the Néel diffusion time (3.32) it is possible to define a blocking Volume V_{Block}^{spm} up to which a particle is superparamagnetic

$$V_{Block}^{spm} = ln(\frac{\tau_m}{\tau_0})\frac{k_B T}{K} \, , \, with \quad T = const \tag{3.33}$$

and vice versa a blocking temperature

$$T_{Block}^{spm} = \frac{KV}{k_B}ln(\frac{\tau_m}{\tau_0}) \qquad V = const \quad . \tag{3.34}$$

This effect is also used to measure the particle size distribution [19].

3.3.3 Combined Rotation

The Néel and Brown like relaxation processes will interact and the shortest time will dominate.

Shliomis and Stepanov showed that the two relaxation processes decouple within linear response theory and that the effective diffusion time τ_{eff} may be calculated by [48]

$$\tau_{eff} = \frac{\tau_B \tau_N}{\tau_B + \tau_N} \quad . \tag{3.35}$$

This linear approximation is not applicable to MPI since the nonlinear magnetization response is of interest.

Nevertheless the magnetization behavior of the particles can be categorized into three groups:

- Group 1: Small particles underlie most likely the Néel diffusion process ($\tau_{eff} \approx \tau_N$).

- Group 2: Intermediate particles are subjected to a relaxation process of a combination of Néel and Brown diffusion.

- Group 3: Large particles are blocked (3.33) and only Brown diffusion happens ($\tau_{\text{eff}} \approx \tau_{\text{B}}$).

3.4 Thermodynamic Equilibrium Properties

A ferrofluid sample, which is in thermodynamic equilibrium at a given temperature T and a constant magnetic field H, has to fulfill the Boltzmann statistics. Considering only particles with an uniaxial anisotropy the energy is given by (3.31)

$$E = K_{\text{eff}} V \sin(\psi)^2 - \mu_0 M_s V H \cos(\theta) \quad . \tag{3.36}$$

The energy is fully determined by the orientation of the magnetic moment $\boldsymbol{m}$ to the magnetic field $\boldsymbol{H}$ and the orientation of the easy axis $\boldsymbol{n}$ to the magnetic moment $\boldsymbol{m}$. Therefore, the state is defined by four variables the polar angles (θ, α) of $\boldsymbol{m}$ and the polar angles (χ, β) of $\boldsymbol{n}$. The geometrical configuration is illustrated in Figure (3.5). The angle ψ between $\boldsymbol{n}$ and $\boldsymbol{m}$ has to be expressed in terms of (θ, α) and (χ, β), which can be done by the following steps:

1. The distance $S1$ shown in Figure (3.6) can be expressed with the help of the cosine formula:

$$S1^2 = \sin(\theta)^2 + \sin(\chi)^2 - 2\sin(\chi)\sin(\theta)\cos(\Phi) \tag{3.37}$$

2. With the additional help of Pythagoras the distance $S2$ is given by:

$$S2^2 = S^1 + (\cos(\chi) - \cos(\theta))^2 \tag{3.38}$$

3. Furthermore $S2$ may also be expressed with the help of the cosine formula:

$$\begin{aligned} S2^2 &= |\boldsymbol{n}|^2 + |\boldsymbol{m}|^2 - 2|\boldsymbol{n}||\boldsymbol{m}|\cos(\psi) \\ &= 2 - 2\cos(\psi) \end{aligned} \tag{3.39}$$

4. Combining the results one finds, that ψ can be expressed in terms of (θ, α) and (χ, β):

$$\cos(\psi) = \cos(\chi)\cos(\theta) + \sin(\chi)\sin(\theta)\cos(\Phi) \tag{3.40}$$

Therefore, the energy can be calculated by

$$E = K_{\text{eff}} V \left(1 - (\cos(\chi)\cos(\theta) + \sin(\chi)\sin(\theta)\cos(\Phi))^2 \right) - \mu_0 M_s V H \cos(\theta) \quad .$$

According to the Boltzmann statistics the probability, that the orientation of the magnetic moment is within the interval $[(\theta, \alpha), (\theta + d\theta, \alpha + d\alpha)]$ and that the easy axis points in the direction of $[(\chi, \beta), (\chi + d\chi, \beta + d\beta)]$ is then given by

$$P(\theta, \alpha, \chi, \beta)\mathrm{d}\beta\mathrm{d}\chi\mathrm{d}\alpha\mathrm{d}\theta = \frac{1}{Z} e^{(\lambda(\cos(\chi)\cos(\theta) + \sin(\chi)\sin(\theta)\cos(\Phi))^2 + \rho\cos(\theta))} \sin(\chi)\sin(\theta)\mathrm{d}\beta\mathrm{d}\chi\mathrm{d}\alpha\mathrm{d}\theta \tag{3.41}$$

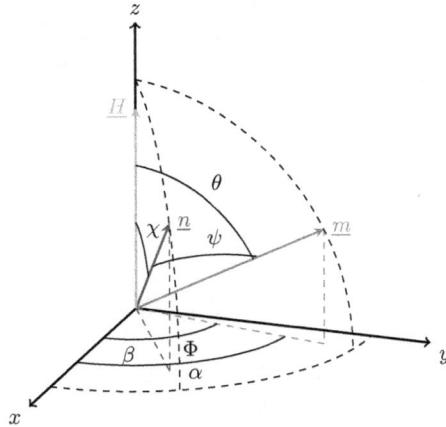

Fig. 3.5: Angles and geometrical configuration used to calculate the thermodynamic equilibrium state. θ, α are the polar angles of the the magnetic moment $\boldsymbol{m}$ and χ, β of the easy axis $\boldsymbol{n}$. ψ is the angle between the magnetic moment and the easy axis and Φ is defined as $\Phi = \alpha - \beta$.

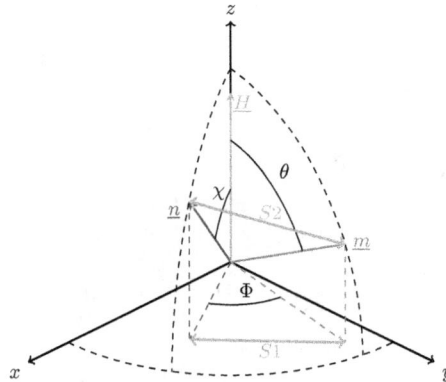

Fig. 3.6: This graphic helps to illustrate, that the angle ψ can be expressed by the polar angles of $\boldsymbol{m}$ and $\boldsymbol{n}$ in terms of $\cos(\psi) = \cos(\chi)\cos(\theta) + \sin(\chi)\sin(\theta)\cos(\Phi)$.

with

$$\rho = \frac{\mu_0 M_s V H}{k_B T} \qquad\qquad \lambda = \frac{K_{eff} V}{k_B T} \quad . \tag{3.42}$$

The partition function Z is therefore given by

$$Z = \int_0^{2\pi} \int_0^{\pi} \int_0^{2\pi} \int_0^{\pi} e^{(\lambda(\cos(\chi)\cos(\theta) + \sin(\chi)\sin(\theta)\cos(\Phi))^2 + \rho\cos(\theta))} \sin(\chi)\sin(\theta) d\beta d\chi d\alpha d\theta \quad . \tag{3.43}$$

By integrating over Φ instead of α, this integral simplifies to

$$Z = 2\pi \int_0^{\pi} e^{\rho\cos(\theta)} \sin(\theta) \left[\int_0^{2\pi} \int_0^{\pi} e^{\lambda(\cos(\chi)\cos(\theta) + \sin(\chi)\sin(\theta)\cos(\Phi))^2} \sin(\chi) d\chi d\Phi \right] d\theta \quad . \tag{3.44}$$

Yasumuri et al [28] showed, that the integral in brackets is constant and independent of θ. Therefore, the first moment of the projection of the magnetic moments on the z axis is given by [10]

$$\langle \cos(\theta) \rangle = \frac{\int_0^{\pi} e^{\rho\cos(\theta)} \cos\theta \sin(\theta) d\theta}{\int_0^{\pi} e^{\rho\cos(\theta)} \sin(\theta) d\theta} \quad . \tag{3.45}$$

This integral can be solved by making use of the substitution $x = \cos(\theta)$.

$$\begin{aligned}
\langle \cos(\theta) \rangle &= \frac{\int_1^{-1} e^{\rho x} x dx}{\int_1^{-1} e^{\rho x} dx} \\
&= \frac{\frac{1}{\rho}(e^{\rho} + e^{\rho}) - \frac{1}{\rho^2}(e^{\rho} - e^{-\rho})}{\frac{1}{\rho}(e^{\rho} - e^{-\rho})} \\
&= \frac{e^{\rho} + e^{\rho}}{e^{\rho} - e^{-\rho}} - \frac{1}{\rho} = \coth(\rho) - \frac{1}{\rho}
\end{aligned} \tag{3.46}$$

The last term is defined as the Langevin function $\mathcal{L}(\rho) = \coth(\rho) - \frac{1}{\rho}$ and furthermore it is possible to calculate the second moment

$$\begin{aligned}
\langle \cos(\theta)^2 \rangle &= \frac{\int_1^{-1} e^{\rho x} x^2 dx}{\int_1^{-1} e^{\rho x} dx} \\
&= 1 + \frac{2}{\rho^2} - \frac{2}{\rho}\coth(\rho) \\
&= 1 - \frac{2}{\rho}\mathcal{L}(\rho) \quad .
\end{aligned} \tag{3.47}$$

Thus, the variance, which can be reagarded as a measure of the thermal fluctuations of the magnetic moments, is given by

$$\langle \cos(\theta)^2 \rangle - \langle \cos(\theta) \rangle^2 = 1 - \frac{2}{\rho}\mathcal{L}(\rho) - (\mathcal{L}(\rho))^2 \quad . \tag{3.48}$$

The result shows, that the equilibrium magnetization does not depend on the anisotropy of the particles but solely on the ratio between the Zeeman energy and the thermal energy.

Therefore, each ferrofluid with a sufficient low density, so that interparticle interactions can be excluded, has the same equilibrium magnetization, which is given by

$$\boldsymbol{M}(\boldsymbol{H}, T) = \frac{N|\boldsymbol{\mu}|}{V_{\mathrm{s}}} \mathcal{L}(\rho) \boldsymbol{e}_H \quad . \tag{3.49}$$

$\boldsymbol{M}(\boldsymbol{H}, T)$ does only depend on the number of particles N, the volume V_{s} of the sample, the magnetic moment of the particles $|\boldsymbol{\mu}|$ and ρ. $\boldsymbol{e}_H$ is the unit vector in the direction of the magnetic field.

The Langevin function is often derived by calculating the classical limit of the Brillouin function. However, the derivation presented here is more reliable, since it includes the anisotropy of the particles.

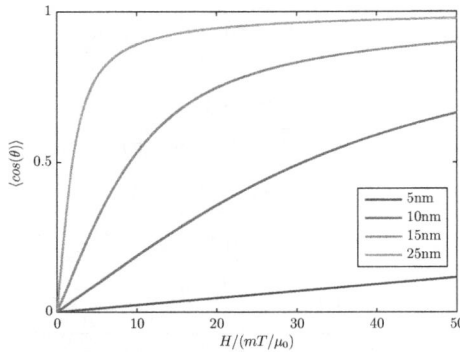

Fig. 3.7: Langevin functions of magnetite nanoparticles at a temperature of 310K for different particle diameters. Obviously larger particles are saturated at far lower magnetic fields than smaller particles.

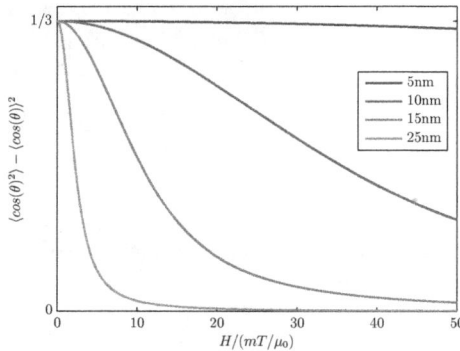

Fig. 3.8: The ordering magnetic force has a drastic influence on large particles, whereas the fluctuations of small particles are still high even at stronger magnetic fields.

4 Stochastic Particle Model

The finite relaxation times present in a ferrofluid (3.3) in combination with a time-varying magnetic field results in a system, which is out of equilibrium. Therefore, the Boltzmann statistics is not any more applicable. However, it is very important for further development of MPI to be able to understand and to simulate the dynamic properties of the contrast agent.

A good approach would be to identify the system as a composition of two subsystems including different dominating time scales. The magnetic particles are orders of magnitude larger, compared to the molecules of the suspension. Thus, in the case of the Brownian rotation the idea of subsystems seems to be applicable [15, 14]. Both subsystems are then linked by the concept of Stokes friction and a stochastic force.

Likewise, the Néel diffusion can be treated as a combination of subsystems, by including a phenomenological damping constant in combination with a random force [20, 67].

By adding a random term to a fully deterministic equation of motion, one ends up with the concept of Langevin equations, which are stochastic differential equations. Dealing with stochastic differential equations some special considerations are needed.

4.1 Stochastic Differential Equation

A simple one-dimensional stochastic differential equations (SDE) is given by

$$\frac{d\xi(t)}{dt} = h(\xi(t), t) + g(\xi(t), t)\lambda(t) \quad , \tag{4.1}$$

where h and g are arbitrary funcitons. h is often called the drift term and g the diffusion term. If the function g also depends on ξ, the noise term $g(\xi(t), t)\lambda(t)$ is called a multiplicative noise term, otherwise an additive noise term.

In general, the time average of the noise is assumed to be zero.

$$\langle \lambda(t) \rangle = 0 \tag{4.2}$$

This is based on the fact, that thermal fluctuation is not creating a directed force, like a deterministic drift force. Furthermore a non-zero mean value could be absorbed into $h(\xi(t), t)$.

The time correlation function of the fluctuation force is often defined by

$$\langle \lambda(t)\lambda(t+\tau)\rangle = \delta(\tau) \quad . \tag{4.3}$$

This delta function takes into account the uncorrelated character of the thermal agitation. For example, each collision of a molecule with a ferrofluid particle is independent of the collisions which happened earlier [34]. Often, an additional prefactor is attached to the delta function, to control the noise intensity and it is called Diffusion factor.

A problem arises due to the non-differentiability of the rapidly fluctuating force and therefore the corresponding integral equation is of importance

$$\xi(t+\tau) - \xi(t) = \int_t^{t+\tau} [h(\xi(t'),t') + g(\xi(t'),t')\lambda(t')]\, \mathrm{d}t' \quad . \tag{4.4}$$

The first part of this integral is of course well defined, whereas the second part needs some additional interpretations. By considering the integral

$$W(t+\Delta t) - W(t) = \int_0^{t+\Delta t} \lambda(t')\mathrm{d}t' \quad , \tag{4.5}$$

one can calculate its mean value

$$\langle W(t+\Delta t) - W(t)\rangle = \int_t^{t+\Delta t} \langle \lambda(t')\rangle \mathrm{d}t' = 0 \tag{4.6}$$

and its variance (see 4.3)

$$\langle (W(t+\Delta t) - W(t))^2\rangle = \int_t^{t+\Delta t} \mathrm{d}s \int_t^{t+\Delta t} \mathrm{d}t' \langle \lambda(s)\lambda(t')\rangle \tag{4.7}$$

$$= \Delta t \quad . \tag{4.8}$$

These results can be used to calculate the drift and diffusion coefficients (see also 4.1.2) [21]

$$D^{(1)} = \lim_{\Delta t \to 0} \frac{\langle W(t+\Delta t) - W(t)\rangle}{\Delta t} = 0 \tag{4.9}$$

$$D^{(2)} = \lim_{\Delta t \to 0} \frac{\langle (W(t+\Delta t) - W(t))^2\rangle}{\Delta t} = 1 \quad , \tag{4.10}$$

$$\tag{4.11}$$

which coincide exactly with the drift and diffusion coefficients within the following Fokker Planck Equation

$$\frac{\partial}{\partial t}P(x,t) = \frac{1}{2}\frac{\partial^2}{\partial x^2}P(x,t) \tag{4.12}$$

of the Wiener process. A stochastic process $W(t)$ is called a Wiener process, if it fulfills the following conditions [51]:

1. $W(0) = 0$

2. $\langle W(t) \rangle = 0$

3. $\langle [W(t) - W(t)]^2 \rangle = (t - s)$ for all $0 \le s \le t$

The increments of the Wiener process are independent and Gaussian distributed with a standard deviation of $\sqrt{\Delta t}$ (4.8) [75]

$$[W(t + \Delta t) - W(t)] \sim \sqrt{\Delta t} \mathcal{N}(0, 1) \quad . \tag{4.13}$$

Here $\mathcal{N}(0, 1)$ denotes a Gaussian random number with standard deviation of 1. The normal distribution of $[W(t + \Delta t) - W(t)]$ arises also very natural in the context of the central limit theorem, since the integral (4.5) can be understood as the sum of an infinite number of random events each with a vanishing mean value.

According to the notation [21]

$$dW = W(t + dt) - W(t) = \lambda(t)dt \quad , \tag{4.14}$$

the second integral term of (4.4) can be interpreted as a Riemann-Stieltjes integral with a Wiener Process as sample function

$$\xi(t + \tau) - \xi(t) = \int_t^{t+\tau} h(\xi(t'), t')dt' + \int_t^{t+\tau} g(\xi(t'), t')dW(t') \quad . \tag{4.15}$$

It is interesting to note, that it is not important to make any assumption about the probability distribution of λ (it is often assumed to be normal distributed [75, 62, 65]). The only properties, which are required are (4.3), (4.2) and that the stochastic process is a continuous process [21].

4.1.1 Ito - Stratonovich Dilemma

Unfortunately, even the integral interpretation of an SDE is not yet well defined. For example the integral

$$\int_{t_0}^{t_e} W(t)dW(t) \tag{4.16}$$

can be calculated as a limit of partial sums [21]

$$S_n = \sum_{i=1}^{N} W(\tau_i)[W(t_i) - W(t_{i-1}))] \quad . \tag{4.17}$$

Here the interval $[t_0, t_e]$ is divided into N subintervals, which are equal in length

$$t_0 < t_1 < t_2 < \cdots < t_e \tag{4.18}$$

and τ_i has to be within the interval $[t_i, t_{i-1}]$

$$\tau_i = \alpha t_i + (1 - \alpha)t_{i-1} \quad \text{with} \quad 0 < \alpha < 1 \quad . \tag{4.19}$$

One can now calculate the mean of (4.17)

$$\langle S_n \rangle = \sum_{i=1}^{N} \langle W(\tau_i)[W(t_i) - W(t_{i-1})] \rangle \tag{4.20}$$

$$= \sum_{i=1}^{N} [\langle W(\tau_i)W(t_i) \rangle - \langle W(\tau_i)W(t_{i-1}) \rangle \tag{4.21}$$

$$= \alpha(t - t_0) \quad , \tag{4.22}$$

by using the identity (here $\tau_i < t_i$)

$$\langle W(\tau_i)W(t_i) \rangle = \langle (W(\tau_i)([W(t_i) - W(\tau_i)] + W(\tau_i)) \rangle \tag{4.23}$$

$$= \langle W(\tau_i)^2 \rangle + \langle W(\tau_i) \rangle \langle (W(t_i) - W(\tau_i)) \rangle \tag{4.24}$$

$$= \tau_i \quad . \tag{4.25}$$

Here it becomes obvious, that a stochastic integral depends on α, which determines the point, where the integral function is evaluated. This is a big difference to non-stochastic integrals, because there it does not matter, since the limit of the upper Darboux sum and the limit of the lower Darboux sum are equal.

This leads to two different calculi. The Itô interpretation evaluates the Integral at the beginning of the time interval

$$\int_{t_0}^{t} g(\xi(t'), t') \mathrm{d}W(t') = \lim_{n \to \infty} \sum_{i=0}^{n} g(\xi(t_i), t_i)[W(t_{i+1}) - W(t_i)] \quad , \tag{4.26}$$

whereas the Stratonovich interpretation averages the stochastic variable $\xi(t)$ over the interval $[t_{i+1}, t_i]$

$$\int_{t_0}^{t} g(\xi(t'), t') \mathrm{d}W(t') = \lim_{n \to \infty} \sum_{i=0}^{n} g\left(\frac{\xi(t_i) + \xi(t_{i+1})}{2}, t_i\right)[W(t_{i+1}) - W(t_i)] \quad . \tag{4.27}$$

Both integral definitions coincide in the case of a simple additive noise term.

However, in the case of a multiplicative noise term it is not clear, which interpretation should be applied, since both definition are mathematical correct, but the results are different.

Fortunately one can switch to the preferred calculus by a transformation. A SDE

$$\frac{\mathrm{d}\xi(t)}{\mathrm{d}t} = h(\xi(t), t) + g(\xi(t), t)\lambda(t) \quad , \tag{4.28}$$

which has to be interpreted in the sense of Itô, may be transformed to a corresponding Stratonovich SDE by subtracting the so called "spurious drift term" $\frac{1}{2}g(\xi(t), t)\frac{\partial}{\partial \xi}g(\xi(t), t)$

$$\frac{\mathrm{d}\xi(t)}{\mathrm{d}t} = h(\xi(t), t) - \frac{1}{2}g(\xi(t), t)\frac{\partial}{\partial \xi}g(\xi(t), t) + g(\xi(t), t) \circ \lambda(t) \quad . \tag{4.29}$$

It is common use to indicate, that the Stratonovich integral definition should be applied by using the symbol "∘". Likewise, the Stratonovich SDE

$$\frac{\mathrm{d}\xi(t)}{\mathrm{d}t} = h(\xi(t), t) + g(\xi(t), t) \circ \lambda(t) \tag{4.30}$$

can be transformed to an Itô SDE

$$\frac{\mathrm{d}\xi(t)}{\mathrm{d}t} = h(\xi(t), t) + \frac{1}{2}g(\xi(t), t)\frac{\partial}{\partial\xi}g(\xi(t), t) + g(\xi(t), t)\lambda(t) \quad . \tag{4.31}$$

The regarded problem determines which calculus should be applied. The assumption, that a process can be described by a white noise process, is often an idealization of a colored noise process with a very short autocorrelation time [34]. If one assumes, that the noise γ is colored noise with an ordinary autocorrelation function

$$\langle\gamma(t+\tau)\gamma(t)\rangle = f_\epsilon(\tau) \tag{4.32}$$

with

$$\lim_{\epsilon \to 0} f_\epsilon(\tau) = \delta(\tau) \quad , \tag{4.33}$$

the Stratonovich interpretation arises naturally in the limit of $\epsilon \to 0$ [62, 21]. This has been proven by Wong and Zakai [16].

In the case of Brown and Néel diffusion processes the noise is caused by thermal fluctuations, which is in fact colored noise and therefore the Stratonovich interpretation should be applied [75].

Another important difference is caused by the applicability of the chain rule. If $f : \mathbb{R} \to \mathbb{R}$ is a twice continuously differentiable function, then the total derivative $\mathrm{d}f(\xi, t)$ in Itô interpretation is given by

$$\mathrm{d}f(\xi, t) = \left(h(\xi, t)\frac{\partial f(\xi, t)}{\partial\xi} + \frac{\partial f(\xi, t)}{\partial t} + g(\xi, t)^2\frac{1}{2}\frac{\partial^2 f(\xi, t)}{\partial\xi^2}\right)\mathrm{d}t + g(\xi, t)\frac{\partial f(\xi, t)}{\partial\xi}\mathrm{d}W \quad . \tag{4.34}$$

Whereas in the case of Stratonovich interpretation the ordinary chain rule can be applied

$$\mathrm{d}f(\xi, t) = \left(h(\xi, t)\frac{\partial f(\xi, t)}{\partial\xi} + \frac{\partial f(\xi, t)}{\partial t}\right)\mathrm{d}t + g(\xi, t)\frac{\partial f(\xi, t)}{\partial\xi}\mathrm{d}W \quad . \tag{4.35}$$

The Itô interpretation is often preferred by mathematicians, because an Itô SDE is a martingale.

4.1.2 Fokker Planck Equation

Instead of solving the Langevin equation and averaging over many trajectories, one can describe the system with a probability distribution $P(\boldsymbol{x}, t)$, whose time evolution follows the Fokker-Planck Equation [62, 75].

$$\frac{\partial P(\boldsymbol{x}, t)}{\partial t} = -\sum_{i=1}^{3}\frac{\partial}{\partial x_i}D_i^{(1)}P(\boldsymbol{x}, t) + \frac{1}{2}\sum_{i,j}\frac{\partial^2}{\partial x_i\partial x_j}D_{ij}^{(2)}P(\boldsymbol{x}, t) \tag{4.36}$$

The variable $\boldsymbol{x}$ incorporates all possible realization of the underlying stochastic process $\boldsymbol{\xi}(t)$. The Langevin equation in three dimensions is given by

$$\frac{\mathrm{d}}{\mathrm{d}t}\boldsymbol{\xi} = \boldsymbol{h}(\boldsymbol{\xi}(t),t) + \hat{\boldsymbol{g}}(\boldsymbol{\xi}(t),t)\boldsymbol{\lambda}(t) \tag{4.37}$$

with the noise term

$$\langle \boldsymbol{\lambda}(t) \rangle = 0 \tag{4.38}$$
$$\langle \lambda_i(t+\tau)\lambda_j(t) \rangle = 2\mathcal{D}\delta_{ij}\delta(\tau) \quad . \tag{4.39}$$

Here $i,j = \{1,2,3\}$ are the indices of Cartesian coordinates, $\mathcal{D}$ is the diffusion constant, which measures the noise, and δ_{ij} is Kronecker's delta

$$\delta_{ij} = \begin{cases} 1 & i = j \\ 0 & i \neq j \end{cases} \quad . \tag{4.40}$$

If (4.37) is integrated in the sense of Stratonovich, the Drift $D_i^{(1)}$ and Diffusion $D_{ij}^{(2)}$ coefficients can be calculated by the relations

$$D_i^{(1)} = h_i + \mathcal{D}\sum_{j,k} g_{jk}\frac{\partial g_{ik}}{\partial x_j} \tag{4.41}$$

$$D_{ij}^{(2)} = 2\mathcal{D}\sum_{k} g_{ik}g_{jk} \quad . \tag{4.42}$$

Therefore, the Drift $D_i^{(1)}$ and Diffusion $D_{ij}^{(2)}$ coefficients can be regarded as the link between the Langevin and the Fokker-Planck approach.

Combining (4.41),(4.42) and (4.36) one can write:

$$\frac{\partial P(\boldsymbol{x},t)}{\partial t} = -\sum_{i=1}^{3}\frac{\partial}{\partial x_i}\left(h_i + \mathcal{D}\sum_{j,k} g_{jk}\frac{\partial g_{ik}}{\partial x_j}\right)P(\boldsymbol{x},t) + \mathcal{D}\sum_{i,j}\frac{\partial^2}{\partial x_i\partial x_j}\left(\sum_{k} g_{ik}g_{jk}\right)P(\boldsymbol{x},t) \tag{4.43}$$

A different notation of the Fokker-Planck equation shows, that it can be understood as a continuity equation.

$$\frac{\partial P(\boldsymbol{x},t)}{\partial t} = -\sum_{i=1}^{3}\frac{\partial}{\partial x_i}\left(\left[h_i - \mathcal{D}\sum_{k} g_{ik}\left(\sum_{j}\frac{\partial g_{jk}}{\partial x_j}\right) - \mathcal{D}\sum_{jk} g_{ik}g_{jk}\frac{\partial}{\partial x_j}\right]P(\boldsymbol{x},t)\right) \tag{4.44}$$

4.1.3 Numerical Integration of SDE

Unfortunately, only very few SDEs can be solved analytically and therefore one has to rely on numerical approximations. The well-known theory of numerical integration of ordinary differential equations cannot be simply transferred to SDEs, due to the non-differentiability of the Wiener process. The book "Numerical Solution of Stochastic Differential Equation" by P.E. Kloeden and E. Platen [51] can be regarded as a reference for numerical integration of SDEs.

There are two criteria of convergence established to evaluate time discrete approximations. The strong convergence criterion is motivated by the definition of the absolute error $\mathcal{E}_s(\Delta_N)$ in terms of

$$\mathcal{E}_s(\Delta_N) = \langle |\xi(T) - X_{\Delta_N}(T)| \rangle \quad . \tag{4.45}$$

Here $\xi(T)$ is the exact solution of the underlying SDE at time T and $X_{\Delta_N}(T)$ is the numerical approximation at the time T, which is obtained by an integration scheme with N equidistant time integration steps $\Delta_N = T/N$. A time discrete approximation X_{Δ_N} is then said to converge strongly with $\gamma > 0$ at time T if there exists a positive constant K, which does not depend on Δ_N, and a $\Delta_0 > 0$ such that

$$\mathcal{E}_s(\Delta_N) = \langle |\xi(T) - X_{\Delta_N}(T)| \rangle \leq K\Delta_N^\gamma \quad \text{with} \quad \Delta_N \in \{0, \Delta_0\} \quad . \tag{4.46}$$

On the other hand the weak convergence criterion says that X_{Δ_N} converges weakly with order $\beta > 0$ at time T, if for each polynomial g there exists a positive constant K, which does not depend on Δ_N, and a $\Delta_0 > 0$ such that

$$\mathcal{E}_w(\Delta_N) = |\langle g(\xi(T)) \rangle - \langle g(X_{\Delta_N}(T)) \rangle| \leq K\Delta_N^\beta \quad \text{with} \quad \Delta_N \in \{0, \Delta_0\} \quad . \tag{4.47}$$

The strong convergence criterion may therefore be understood as a pathwise convergence criterion, whereas the weak order convergence criterion judges the approximation of the probability distribution. Weak and strong order time approximation schemes differ from each other. The weak approximation schemes are often simpler and it is easier to obtain a higher order of convergence. Furthermore they are often faster, because one is not restricted to use Gaussian random number generators [44]. Since the main aim of this thesis is to generate the first statistical moment of the magnetic moments, one could think of implementing weak order approximations. This is misleading insofar, as the combined rotation relies on the connection of the Brownian rotation and the Néel rotation, which is only possible if the detailed solution paths are known.

The general derivation of numerical integration schemes is based on the stochastic Taylor expansions. Here, one has to be careful, because Itô and Stratonovich Taylor expansions are different, due to the different chain rules (4.35, 4.34). Since the Stratonovich interpretation should be applied, only the Stratonovich Taylor expansions are considered. Applying the Stratonovich chain rule (4.35) to the Stratonovich SDE

$$\xi(t) = \xi(t_0) + \int_{t_0}^t h(\xi(s))\mathrm{d}s + \int_{t_0}^t g(\xi(s)) \circ \mathrm{d}W(s) \tag{4.48}$$

one gets

$$\begin{aligned} f(\xi(t)) &= f(\xi(t_0)) + \int_{t_0}^t h(\xi(s))\frac{\partial f}{\partial \xi}\mathrm{d}s + \int_{t_0}^t g(\xi(s))\frac{\partial f}{\partial \xi} \circ \mathrm{d}W(s) \\ &= f(\xi(t_0)) + \int_{t_0}^t L^0 f(\xi(s))\mathrm{d}s + \int_{t_0}^t L^1 f(\xi(s)) \circ \mathrm{d}W(s) \quad , \end{aligned} \tag{4.49}$$

here the operators

$$L^0 = h(\xi)\frac{\partial}{\partial \xi} \qquad (4.50)$$

$$L^1 = g(\xi)\frac{\partial}{\partial \xi} \qquad (4.51)$$

have been used. The first term of the Taylor expansion may then be derived by applying the chain rule (4.49) to the functions $f(\xi) = h(\xi)$, $f(\xi) = g(\xi)$ and putting the results back into (4.48)

$$
\begin{aligned}
\xi(t) =& \xi(t_0) + \int_{t_0}^t \left(h(\xi(t_0)) + \int_{t_0}^s L^0 h(\xi(z))\mathrm{d}z + \int_{t_0}^s L^1 h(\xi(z)) \circ \mathrm{d}W(z) \right) \mathrm{d}s \\
&+ \int_{t_0}^t \left(g(\xi(t_0)) + \int_{t_0}^s L^0 g(\xi(z))\mathrm{d}z + \int_{t_0}^s L^1 g(\xi(z)) \circ \mathrm{d}W(z) \right) \circ \mathrm{d}W(s) \quad (4.52) \\
=& \xi(t_0) + h(\xi(t_0)) \int_{t_0}^s \mathrm{d}s + g(\xi(t_0)) \int_{t_0}^s \circ \mathrm{d}W(s) + \mathcal{R}_1 \quad .
\end{aligned}
$$

This is actually the basis of the simple Euler-Maruyama integration scheme. $\mathcal{R}$ is called the reminder and includes all integrals with non constant integrands. One then may repeat this procedure with $f = L^1 g(\xi)$ to get the next Taylor step

$$
\begin{aligned}
\xi(t) =& \xi(t_0) + h(\xi(t_0)) \int_{t_0}^s \mathrm{d}s + g(\xi(t_0)) \int_{t_0}^s \circ \mathrm{d}W(s) \\
&+ L^1 g(\xi(t_0)) \int_{t_0}^t \int_{t_0}^s \circ \mathrm{d}W(z) \circ \mathrm{d}W(s) + \mathcal{R}_2 ,
\end{aligned} \qquad (4.53)
$$

whereas the remainder is given by:

$$
\begin{aligned}
\mathcal{R}_2 =& \int_{t_0}^t \int_{t_0}^s L^0 h(\xi(z))\mathrm{d}z\mathrm{d}s + \int_{t_0}^t \int_{t_0}^s L^1 h(\xi(z))\mathrm{d}W(z)\mathrm{d}s \\
&+ \int_{t_0}^t \int_{t_0}^s L^0 g(\xi(z))\mathrm{d}z\mathrm{d}W(s) + \int_{t_0}^t \int_{t_0}^s \int_{t_0}^z L^0 L^1 g(\xi(u))\mathrm{d}u\mathrm{d}W(z)\mathrm{d}W(s) \\
&+ \int_{t_0}^t \int_{t_0}^s \int_{t_0}^z L^1 L^1 g(\xi(u))\mathrm{d}W(u)\mathrm{d}W(z)\mathrm{d}W(s)
\end{aligned}
$$

The equation (4.53) leads to the Milstein integration scheme. Higher order Taylor expansion include multiple stochastic integrals, which can be approximated as presented in [51]. One has to be careful with adopting the integration schemes from [51], since they are only valid in case of autonomous differential equation. The Brown and Néel diffusion processes are of course non-autonomous differential equations because of the magnetic field, nevertheless the expansions can be adopted up to the Milstein scheme, since the time-dependent magnetic field is only present in the drift coefficient. In this thesis the Euler-Maruyama and the Heun scheme have been implemented, which are both well established.

Euler-Maruyama

The integration step for a equidistant time discretization $0 < \Delta_N < 2\Delta_N < \cdots < T$ of the time interval $[0, T]$ with $\Delta_N = T/N$ is given by

$$X_k(t + \Delta_N) = X_k(t) + h_k(\boldsymbol{X}(t), t)\Delta_N + \sum_{i=1}^{3} g_{ik}(\boldsymbol{X}(t))\Delta W_i \quad . \tag{4.54}$$

$X(t)_k$ denotes the kth component at the time t of the stochastic process and h_k, g_{ik} are the drift and diffusion terms. ΔW_k is the increment of the kth component of the Wiener process. According to (4.13) they are Gaussian random numbers with standard deviation given by $\sqrt{\Delta_N}$.

If one would like to approximate the Stratonovich solution of the underlying stochastic process, one has to augment the drift term with the spurious drift, since the Euler-Maruyama scheme converges naturally in the sense of Itô

$$\begin{aligned} X_k(t + \Delta_N)_k =& X_k(t) + \left(h_k(\boldsymbol{X}(t), t) + \mathcal{D} \sum_{j=1}^{3} \sum_{i=1}^{3} g_{ji}(\boldsymbol{X}(t)) \frac{\partial g_{ki}(\boldsymbol{X}(t))}{\partial X_j} \right) \Delta_N \\ &+ \sum_{i=1}^{3} g_{ik}(\boldsymbol{X}(t))\Delta W_i \quad . \end{aligned} \tag{4.55}$$

The Euler-Maruyama scheme converges with $\gamma = \frac{1}{2}$ according to the strong order convergence and with $\beta = 1$ referred to the weak order convergence criterion.

Heun

The Heun scheme is a predictor-corrector method, which is in extensive use in micromagnetic simulations [6, 20, 50, 72]. The predictor is given by the simple Euler-Maruyama step

$$\overline{X}(t + \Delta_N)_k = X_k(t) + h_k(\boldsymbol{X}(t), t)\Delta_N + \sum_{i=1}^{3} g_{ik}(\boldsymbol{X}(t))\Delta W_i \quad , \tag{4.56}$$

which is then followed by the corrector step

$$\begin{aligned} X_k(t + \Delta_N) =& \frac{1}{2} \left(h_k(\overline{\boldsymbol{X}}(t + \Delta_N), t + \Delta_N) + h_k(\boldsymbol{X}(t), t) \right) \Delta_N \\ &+ \frac{1}{2} \sum_{i=1}^{3} \left(g_{ik}(\overline{\boldsymbol{X}}(t + \Delta_N)) + g_{ik}(\boldsymbol{X}(t)) \right) \Delta W_i \quad . \end{aligned} \tag{4.57}$$

The Heun integration scheme incorporates implicitness, which should enhance its stability. It converges directly to the Stratonovich interpretation [61].

Milstein and higher order schemes

According to (4.53), the next higher order integration scheme is given by

$$
\begin{aligned}
X_k(t + \Delta_N) = & X_k(t) + h_k(\boldsymbol{X}(t), t)\Delta_N + \sum_{i=1}^{3} g_{ik}(\boldsymbol{X}(t))\Delta W_i \\
& + \sum_{j=1}^{3}\sum_{i=1}^{3}\sum_{l=1}^{3} g_{lj}(\boldsymbol{X}(t))\frac{\partial g_{ki}(\boldsymbol{X}(t))}{\partial X_l}\mathcal{J}_{(j,i)} \quad .
\end{aligned}
\tag{4.58}
$$

This is the so-called Milstein scheme, which achieves a strong order convergence of $\gamma = 1$. Here $\mathcal{J}_{(j,i)}$ denotes the stochastic double integral

$$
\mathcal{J}_{(j,i)} = \int_{n\Delta_N}^{(n+1)\Delta_N} \int_{n\Delta_N}^{s} \circ dW_j(z) \circ dW_i(s) \quad .
\tag{4.59}
$$

Unfortunately, in general this integral cannot be expressed by ΔW_j and ΔW_i and therefore it has to be approximated. This requires additional random numbers and thus it is very time-consuming [64]. The integral $\mathcal{J}_{(j,i)}$ can only be easily calculated if the diffusion coefficients satisfy the following commutativity conditions:

$$
\sum_{l=1}^{3} g_{lj}(\boldsymbol{X}(t))\frac{\partial g_{ki}(\boldsymbol{X}(t))}{\partial X_l} = \sum_{l=1}^{3} g_{li}(\boldsymbol{X}(t))\frac{\partial g_{kj}(\boldsymbol{X}(t))}{\partial X_l}
\tag{4.60}
$$

However, in the special case of Néel or Brown diffusion processes this condition is not valid. Therefore, higher order of accuracy would increase the computational effort dramatically, whereas the gain in numerical approximation is small compared to the present statistical error of the mean of N solution paths, since its variance decreases only with $\propto 1/\sqrt{N}$ [64].

Implementation

The Euler-Maruyama and the Heun scheme have been implemented in C++. The required random numbers are generated by the pseudorandom number generator Mersenne Twister, MT19937, which is implemented in the gnu scientific library version 1.14 [46]. The MT19937 has a very long period of $2^{19937} - 1$ and accomplishes numerous tests for statistical randomness [47].

The uniformly distributed random numbers are used to generate independent standard normally distributed random numbers. This transformation is accomplished by the Ziggurat algorithm, which is also implemented in the gnu scientific library.

The C++ program is controlled by a matlab script, whereby text files are used to exchange data. Matlab has also been used to analyze the simulation data.

4.2 Brownian Rotation

The equation of motion of a sphere with a fixed dipole moment $\boldsymbol{\mu}$ is given by [75]

$$\underbrace{\hat{\boldsymbol{I}}\dot{\boldsymbol{\omega}}}_{Inertia} = \underbrace{\boldsymbol{\mu} \times \mu_0 \boldsymbol{H}}_{Torque} + \underbrace{\boldsymbol{\lambda}(t)}_{Noise} - \underbrace{\hat{\boldsymbol{\zeta}}\boldsymbol{\omega}}_{Friction} \quad , \tag{4.61}$$

where the noise term is characterized by

$$\langle \boldsymbol{\lambda}(t) \rangle = 0$$
$$\langle \lambda_i(t + \tau)\lambda_j(t) \rangle = 2\mathcal{D}_{\mathrm{B}}\delta_{ij}\delta(\tau) \quad .$$

If one assumes that the particle is an ideal sphere, the frictional tensor $\hat{\boldsymbol{\zeta}}$ reduces to a scalar ζ. Furthermore, one can neglect all inertial effects, as shown in (3.3.1) and can make use of the relation

$$\frac{\mathrm{d}}{\mathrm{d}t}\boldsymbol{\mu} = \boldsymbol{\omega} \times \boldsymbol{\mu} \tag{4.62}$$

with the angular velocity $\boldsymbol{\omega}$ and the dipole moment $\boldsymbol{\mu}$. This leads to the equation of motion, which characterizes the Brownian rotation

$$\zeta\dot{\boldsymbol{\mu}} = \boldsymbol{\lambda}(t) \times \boldsymbol{\mu} + [\boldsymbol{\mu} \times \mu_0 \boldsymbol{H}] \times \boldsymbol{\mu} \quad . \tag{4.63}$$

The two cross products on the right hand side ensure the conservation of the norm of $\boldsymbol{\mu}$.

Considering the identity $\boldsymbol{a} \times (\boldsymbol{b} \times \boldsymbol{c}) = \boldsymbol{b} \cdot (\boldsymbol{a} \cdot \boldsymbol{c}) - \boldsymbol{c} \cdot (\boldsymbol{a} \cdot \boldsymbol{b})$ and introducing the unit vector $\boldsymbol{m} = \boldsymbol{\mu}/|\boldsymbol{\mu}|$, one can reformulate (4.63) to:

$$\frac{\mathrm{d}}{\mathrm{d}t}\boldsymbol{m} = \frac{\mu_0|\boldsymbol{\mu}|}{\zeta}\left[\boldsymbol{H} - \boldsymbol{m}(\boldsymbol{m}\boldsymbol{H})\right] - \frac{1}{\zeta}\left[\boldsymbol{m} \times \boldsymbol{\lambda}\right] \tag{4.64}$$

One may of course also rearrange (4.63) to

$$\frac{\mathrm{d}}{\mathrm{d}t}\boldsymbol{m} = -\frac{1}{\zeta}\boldsymbol{m} \times \left(\mu_0|\boldsymbol{\mu}|\left[\boldsymbol{m} \times \boldsymbol{H}\right] + \boldsymbol{\lambda}\right) \quad . \tag{4.65}$$

Since the Brownian Rotation conserves the norm, the function $d(\boldsymbol{m})^2/dt$ should be zero. Calculating this in terms of the Stratonovich interpretation, which is easy, because the ordinary chain rule applies, one gets

$$\frac{\mathrm{d}(\boldsymbol{m})^2}{\mathrm{d}t} = 2\boldsymbol{m}\frac{\mathrm{d}\boldsymbol{m}}{\mathrm{d}t} = 2\boldsymbol{m}(\boldsymbol{m} \times (\ldots)) = 0 \quad . \tag{4.66}$$

This is no longer correct, if the Itô calculus would be used. This is an additional argument for applying Stratonovich calculus.

4.2.1 The Fokker Planck Equation

According to (4.44), the Fokker Planck equation of the Brownian Rotation may be derived by first identifying the drift and diffusion functions h_i and g_{ik} of (4.64)

$$h_i = \frac{\mu_0|\boldsymbol{\mu}|}{\zeta}[H_i - m_i(\boldsymbol{mH})] \tag{4.67}$$

$$g_{ik} = -\frac{1}{\zeta}\sum_p \epsilon_{ipk}m_p \quad, \tag{4.68}$$

where ϵ_{ipk} is the Levi-Civitas permutation symbol. It has the following properties:

$$\epsilon_{ipk} = \begin{cases} +1 & \text{if } (ipk) \text{ is an even permutation of } (1,2,3) \\ -1 & \text{if } (ipk) \text{ is an odd permutation of } (1,2,3) \\ 0 & \text{if } i = p \text{ or } i = k \text{ or } p = k \end{cases} \tag{4.69}$$

By calculating

$$\sum_j \frac{\partial g_{jk}}{\partial m_j} = \frac{-1}{\zeta}\sum_j \partial_{m_j}\sum_j \epsilon_{jpk}m_p$$

$$= 0 \tag{4.70}$$

and

$$\sum_{jk} g_{ik}g_{jk}\frac{\partial}{\partial \boldsymbol{m}} = \frac{1}{\zeta^2}\sum_{jk}\left(\sum_l \epsilon_{ilk}m_l\right)\left(\sum_p \epsilon_{jpk}m_p\right)\frac{\partial}{\partial \boldsymbol{m}} \tag{4.71}$$

$$= \frac{1}{\zeta^2}\left[\boldsymbol{m} \times \left(\boldsymbol{m} \times \frac{\partial}{\partial \boldsymbol{m}}\right)\right] \tag{4.72}$$

one can write down the Fokker Planck Equation of the Brownian diffusion

$$\frac{\partial P(\boldsymbol{m},t)}{\partial t} = -\frac{\partial}{\partial \boldsymbol{m}}\left(\frac{\mu_0|\boldsymbol{\mu}|}{\zeta}[\boldsymbol{H} - m(\boldsymbol{mH})] - \frac{\mathcal{D}_{\mathrm{B}}}{\zeta^2}\left[\boldsymbol{m} \times \left(\boldsymbol{m} \times \frac{\partial}{\partial \boldsymbol{m}}\right)\right]\right)P(\boldsymbol{m},t). \tag{4.73}$$

The still unknown diffusion constant $\mathcal{D}_{\mathrm{B}}$ can be specified by comparing the probability density $P(\boldsymbol{m},t)$ to the Boltzmann distribution. If the system is in equilibrium state, the time differential vanishes

$$\frac{\partial P(\boldsymbol{m},t)}{\partial t} = 0$$

and the probability density $P(\boldsymbol{m},t)$ has to be equal to the Boltzmann distribution [75]. The Boltzmann distribution of the Brownian rotation is solely determined by the ratio of the Zeeman to the thermal energy (3.4)

$$P_0(\boldsymbol{m},t) = A\exp\left(\frac{\mu_0|\boldsymbol{\mu}|\boldsymbol{mH}}{k_{\mathrm{B}}T}\right) \quad, \tag{4.74}$$

where A is a normalization factor. Inserting (4.74) into (4.73) one gets

$$0 = -\frac{\partial}{\partial \boldsymbol{m}} \left(\frac{\mu_0 |\boldsymbol{\mu}|}{\zeta} [\boldsymbol{H} - \boldsymbol{m}(\boldsymbol{mH})] - \frac{\mu_0 |\boldsymbol{\mu}| \mathcal{D}_B}{\zeta^2 k_B T} [\boldsymbol{H} - \boldsymbol{m}(\boldsymbol{mH})] \right) P_0(\boldsymbol{m}, t) . \quad (4.75)$$

It follows, that the diffusion constant $\mathcal{D}_B$ has to be equal to the friction times the thermal energy

$$\mathcal{D}_B = \zeta k_B T \quad . \quad (4.76)$$

4.2.2 Solution of the Fokker Planck Equation

Of course one may think of solving the Fokker Planck Equation (4.73) directly. This would have the advantage, that the Fokker Planck solution would contain no noise. Due to the radial symmetry of the problem the Fokker Planck Equation is transformed to spherical coordinates.

$$2\tau_B \frac{\partial P(\theta, \phi)}{\partial t} = \Delta P(\theta, \phi) + \frac{1}{k_B T} \left[\frac{1}{\sin(\theta)} \frac{\partial}{\partial \theta} \left(\sin(\theta) P(\theta, \phi) \frac{\partial V}{\partial \theta} \right) \right.$$
$$\left. + \frac{1}{\sin^2(\theta)} \frac{\partial}{\partial \phi} \left(P(\theta, \phi) \frac{\partial V}{\partial \phi} \right) \right] \quad (4.77)$$

Here

$$\Delta = \frac{1}{\sin(\theta)} \frac{\partial}{\partial \theta} \left(\sin(\theta) \frac{\partial}{\partial \theta} \right) + \frac{1}{\sin(\theta)^2} \frac{\partial^2}{\partial \phi^2} \quad (4.78)$$

is the Laplace Operator in spherical coordinates and τ_B is given by (3.26). This equation simplifies, if one assumes, that the magnetic field is unidirectional, so that the potential is given by

$$V = -\mu_0 |\boldsymbol{\mu}| H \cos(\theta) \quad . \quad (4.79)$$

In this case, the Fokker Planck Equation no longer depends on ϕ:

$$2\tau_B \frac{\partial P(\theta, t)}{\partial t} = \frac{1}{\sin(\theta)} \frac{\partial}{\partial \theta} \left[\sin(\theta) \left(\frac{\partial P(\theta, t)}{\partial \theta} + \frac{\mu_0 |\boldsymbol{\mu}| H}{k_B T} \sin(\theta) P(\theta, t) \right) \right] \quad (4.80)$$

This one dimensional Fokker Planck Equation can then be solved by an expansion in Legendre polynomials

$$P(\theta, t) = \sum_{n=0}^{\infty} a_n(t) L_n(\cos(\theta)) \quad . \quad (4.81)$$

By substituting $x = cos(\theta)$ and inserting the ansatz (4.81) into (4.80) one gets

$$2\tau_B \sum_{n=0}^{\infty} P_n(x) \partial_t a_n(t) = \sum_{n=0}^{\infty} a_n(t) \frac{d}{dx} \left[(1 - x^2) \left(\frac{d}{dx} L_n(x) - \rho L_n(x) \right) \right] \quad , (4.82)$$

where $\rho = \frac{\mu_0 |\mu| H}{k_B T}$ has been introduced. The right term within the sum can be rearranged to:

$$\underbrace{\frac{d}{dx}\left[(1 - x^2)\frac{d}{dx}L_n\right]}_{1} + \underbrace{\left[2\rho x L_n + (x^2 - 1)\rho\frac{d}{dx}L_n\right]}_{2} \tag{4.83}$$

According to the recurrence formula

$$(x^2 - 1)\frac{d}{dx}L_n(x) = n\left[xL_n(x) - L_{n-1}(x)\right] \qquad \text{for} \quad n \geq 1 \tag{4.84}$$

of Legendre polynomials, the first part can be written as

$$n\left[\frac{d}{dx}L_{n-1} - x\frac{d}{dx}L_n(x) - L_n\right] \quad . \tag{4.85}$$

The recurrence formula (4.84) may be also written in terms of

$$\frac{d}{dx}L_n(x) = \frac{(n)\left[xL_n(x) - L_{n-1}\right]}{(x^2 - 1)} \tag{4.86}$$

or

$$\frac{d}{dx}L_{n-1}(x) = \frac{(n - 1)\left[xL_{n-1}(x) - L_{n-2}\right]}{(x^2 - 1)} \quad , \tag{4.87}$$

if a shift $n \to n - 1$ is performed. According to (4.86) and (4.87), one can now change (4.85) to:

$$n\left[\frac{(2n - 1)xL_{n-1}(x) - nx^2L_n(x) - (n - 1)L_{n-2}(x)}{(x^2 - 1)} - L_n(x)\right] \tag{4.88}$$

Considering a different definition of the recurrence formula

$$(n + 1)L_{n+1}(x) = (2n + 1)xL_n(x) - nL_{n-1}(x) \tag{4.89}$$

and performing an additional shift $n \to n - 1$ one gets

$$n\left[\frac{nL_n(x) + (n - 1)L_{n-2} - nx^2L_n(x) - (n - 1)L_{n-2}}{(x^2 - 1)} - L_n(x)\right] \quad , \tag{4.90}$$

which is equal to:

$$-nL_n(x)(n + 1) \qquad \text{with} \quad n \geq 2 \tag{4.91}$$

To rearrange the second term of (4.83), one first needs the recurrence formula (4.84)

$$\left[2\rho x L_n(x) + n\rho\left(xL_n(x) - L_{n-1}(x)\right)\right] \quad . \tag{4.92}$$

Furthermore, (4.89) may be used to transfer this to

$$\rho \left[\frac{2+n}{2n+1} \left[(n+1)L_{n+1}(x) + nL_{n-1} \right] - nL_{n-1}(x) \right] \tag{4.93}$$

and finally to

$$\rho \left[nL_{n-1}(x) \frac{1-n}{2n+1} + L_{n+1}(x) \frac{(2+n)(n+1)}{(2n+1)} \right] \quad . \tag{4.94}$$

Thus (4.82) is equivalent to

$$2\tau_{\mathrm{B}} \sum_{n=2}^{\infty} \partial_t a_n(t) L_n(x)$$

$$= \sum_{n=2}^{\infty} a_n(t) \left\{ \rho \left[nL_{n-1}(x) \frac{1-n}{2n+1} + L_{n+1}(x) \frac{(2+n)(n+1)}{(2n+1)} \right] - nL_n(x)(n+1) \right\} \quad . \tag{4.95}$$

Now, it is possible to take advantage of the orthogonality-relationship of the Legendre polynomials. The orthogonality-relationship is given by

$$\int_{-1}^{1} L_n(x) L_m(x) \mathrm{d}x = \begin{cases} 0 & n \neq m \\ \frac{2}{(2m+1)} & n = m \end{cases} \tag{4.96}$$

and therefore the following relation is valid:

$$\sum_{n=0}^{\infty} f_n \int_{-1}^{1} L_{n-1} L_m \mathrm{d}x = f_{n+1} \frac{2}{(2m+1)} \tag{4.97}$$

By multiplying (4.95) with L_n and integrating it over x, one finally ends up with a differential recurrence equation

$$\frac{2\tau_{\mathrm{B}}}{n(n+1)} \partial_t a_n(t) = \rho \left(\frac{a_{n-1}(t)}{2n-1} - \frac{a_{n+1}(t)}{2n+3} \right) - a_n(t) \quad . \tag{4.98}$$

Since this equation is only valid for $n \geq 2$ one has to individually consider $n = 0$ and $n = 1$. To do so, one has to calculate the following terms:

$$\frac{\mathrm{d}}{\mathrm{d}x} \left[(1-x^2) \frac{\mathrm{d}}{\mathrm{d}x} L_0 \right] + \left[2\rho x L_0 + (x^2-1)\rho \frac{\mathrm{d}}{\mathrm{d}x} L_0 \right] = 2L_1 \rho$$

$$\frac{\mathrm{d}}{\mathrm{d}x} \left[(1-x^2) \frac{\mathrm{d}}{\mathrm{d}x} L_1 \right] + \left[2\rho x L_1 + (x^2-1)\rho \frac{\mathrm{d}}{\mathrm{d}x} L_1 \right] = 2L_2 \rho - 2L_1 \tag{4.99}$$

$$\frac{\mathrm{d}}{\mathrm{d}x} \left[(1-x^2) \frac{\mathrm{d}}{\mathrm{d}x} L_2 \right] + \left[2\rho x L_2 + (x^2-1)\rho \frac{\mathrm{d}}{\mathrm{d}x} L_2 \right] = -\frac{2}{3} L_2 + \rho \left[\frac{12}{5} L_3 - \frac{2}{5} L_1 \right]$$

These are necessary to initialize (4.82)

$$2\tau_B\partial_t\left[a_0(t)L_0 + a_1(t)L_1 + a_2(t)L_2 + \ldots\right] =$$
$$a_0(t)2L_1\xi + a_1(t)2L_2\xi - 2L_1 + a_2(t)\left(\frac{-2}{3}L_2 + \xi\left[\frac{12}{5}L_3 - \frac{2}{5}L_1\right]\right)\ldots \tag{4.100}$$

By again taking advantage of the orthogonality-relationship one finally obtains the solution

$$\text{for} \quad n = 0 \tag{4.101}$$
$$\partial_t a_0(t) = 0$$
$$\text{for} \quad n \geq 1$$
$$\frac{2\tau_B}{n(n+1)}\partial_t a_n(t) = \rho\left(\frac{a_{n-1}(t)}{2n-1} - \frac{a_{n+1}(t)}{2n+3}\right) - a_n(t) \quad .$$

This solution coincides with that one presented in [76]. Differential recurrence equations can be solved by a method called matrix continued fraction, which is presented in [75]. However, in this case of simple scalars this equation can be solved iteratively.

The derivation of the general solution (not restricted to an unidirectional magnetic field) of the Fokker-Planck equation of the Brownian diffusion is far more complicated. Furthermore, there is no Fokker-Planck equation, which is able to handle the combination of the Néel and Brownian diffusion. Therefore, subsequently only the direct numerical integration of the Langevin equations is applied. Nevertheless, this solution can be used to test the numerical integration schemes.

4.2.3 Numerical Integration

Euler-Maruyama

First, one has to calculate the spurious drift term, since the Euler-Maruyama integration scheme converges in sense of Itô

$$\sum_{j=1}^{3}\sum_{i=1}^{3} g_{ji}\frac{\partial g_{ki}}{\partial m_j} = \frac{1}{\zeta^2}\sum_{j=1}^{3}\sum_{i=1}^{3}\sum_{p=1}^{3}\epsilon_{jpi}m_p\epsilon_{kji}$$
$$= -\frac{1}{\zeta^2}\sum_{p=1}^{3}2\delta_{pk}m_p = -\frac{2}{\zeta^2}m_k \quad . \tag{4.102}$$

The Stokes friction for the rotational motion of a sphere is given by

$$\zeta = 6\eta V_h \quad , \tag{4.103}$$

where η is the viscosity of the surrounding medium. Therefore, according to (4.55), the Euler-Maruyama integration scheme to integrate the Brownian diffusion problem is given by

$$
\begin{aligned}
\boldsymbol{m}(t + \Delta t) = &\boldsymbol{m}(t) + \left(\frac{\mu_0 M_\mathrm{s} V_\mathrm{c}}{6\eta V_\mathrm{h}} \left[\boldsymbol{H}(t) - \boldsymbol{m}(t)(\boldsymbol{m}(t)\boldsymbol{H}(t)) \right] - \frac{2k_\mathrm{B}T}{6\eta V_\mathrm{h}} \boldsymbol{m} \right) \Delta t \\
&- \sqrt{\frac{2k_\mathrm{B}T}{6\eta V_\mathrm{h}}} \boldsymbol{m}(t) \times \Delta \boldsymbol{W}(t) \quad .
\end{aligned}
\tag{4.104}
$$

Heun

The predictor is given by

$$
\overline{\boldsymbol{m}}(t + \Delta t) = \boldsymbol{m}(t) + \frac{\mu_0 M_\mathrm{s} V_\mathrm{c}}{6\eta V_\mathrm{h}} \left[\boldsymbol{H}(t) - \boldsymbol{m}(t)(\boldsymbol{m}(t)\boldsymbol{H}(t)) \right] \Delta t - \sqrt{\frac{2k_\mathrm{B}T}{6\eta V_\mathrm{h}}} \boldsymbol{m}(t) \times \Delta \boldsymbol{W}(t)
\tag{4.105}
$$

and the corrector reads

$$
\begin{aligned}
\boldsymbol{m}(t + \Delta t) = &\boldsymbol{m}(t) \\
&+ \frac{1}{2} \frac{\mu_0 M_\mathrm{s} V_\mathrm{c}}{6\eta V_\mathrm{c}} \left([\boldsymbol{H}(t + \Delta t) - \overline{\boldsymbol{m}}(t + \Delta t)(\overline{\boldsymbol{m}}(t + \Delta t)\boldsymbol{H}(t + \Delta t))] \right. \\
&\qquad\qquad \left. + [\boldsymbol{H}(t) - \boldsymbol{m}(t)(\boldsymbol{m}(t)\boldsymbol{H}(t))] \right) \Delta t \\
&- \frac{1}{2} \sqrt{\frac{2k_\mathrm{B}T}{6\eta V_\mathrm{h}}} \left(\overline{\boldsymbol{m}}(t + \Delta t) + \boldsymbol{m}(t) \right) \times \Delta \boldsymbol{W}(t) \quad .
\end{aligned}
\tag{4.106}
$$

4.2.4 Tests

The two integration schemes, Heun and Euler-Maruyama, can only be tested with respect to the weak order convergence criterion, since there is no known analytical solution of the Brownian diffusion Langevin equation. Nevertheless, it is possible to test the statistical properties of a equilibrated particle ensemble, see Figure 4.1.

If the ensemble is in thermodynamic equilibrium the probability distribution $P_\mathrm{B}(\theta)$ of the angle θ between the reduced magnetic moments on the axis of the applied magnetic field has to coincide with (3.4):

$$
P_\mathrm{B}(\theta) = \frac{1}{Z} \sin(\theta) \exp\left(\frac{\mu_0 V_\mathrm{c} M_\mathrm{s} H_z \cos(\theta)}{k_\mathrm{B} T} \right)
\tag{4.107}
$$

Both integration schemes show good agreement with $P_\mathrm{B}(\theta)$, if a sufficient small time integration step size is applied, see Figure (4.3). It has also been checked, that the final probability distribution does not depend on the friction of the Brownian rotation. Parameters like the viscosity or the hydrodynamic diameter of the particle only have influence on the time the system needs to equilibrate. Furthermore, since the Fokker Planck Equation has been solved, it is possible to compare the solution obtained by solving the differential recurrence equation (4.102) and the numerical results of the SDE.

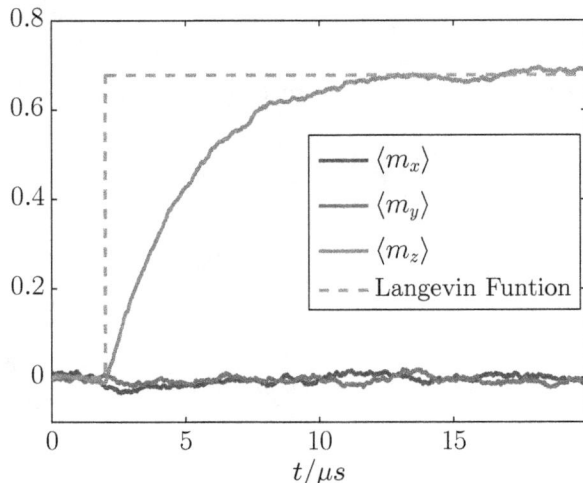

Fig. 4.1: Equilibration of a particle ensemble of 1000 particles with $d_\mathrm{c} = 15\,\mathrm{nm}$ and $d_\mathrm{h} = 25\,\mathrm{nm}$. The magnetic field $H_z = 15\,\mathrm{mT}/\mu_0$ is switched on at $t = 2\,\mu s$.

It has been found that both results coincide perfectly, see figure (4.2).

The time step dependence can be evaluated by comparing the deviation of the analytical predicted mean value (3.46) and variance (3.48) in dependence of the time integration step size Δt. The statistical properties of each integration-run have been evaluated at the same time after the magnetic field has been switched on. Figure (4.4) shows the results. It has been found that the Euler-Maruyama method better suits the Brownian diffusion integration problem and it is also numerically less complex compared to the Heun scheme. Therefore, the Euler-Maruyama method should be used to integrate the Langevin equation of the Brownian diffusion.

4.3 Néel Rotation

A magnetic system, which is in thermodynamic equilibrium, fulfills the condition [41]

$$\boldsymbol{\mu}(\boldsymbol{r}) \times \boldsymbol{H}^{\mathrm{eff}} = 0 \qquad \text{with} \tag{4.108}$$
$$\boldsymbol{H}^{\mathrm{eff}} = \boldsymbol{H}^{\mathrm{ext}} + \boldsymbol{H}^{\mathrm{MS}} + \boldsymbol{H}^{\mathrm{An}} + \boldsymbol{H}^{\mathrm{Ex}} \quad .$$

Here $\boldsymbol{H}^{\mathrm{ext}}$, $\boldsymbol{H}^{\mathrm{MS}}$, $\boldsymbol{H}^{\mathrm{An}}$ and $\boldsymbol{H}^{\mathrm{Ex}}$ are the external magnetic field, the magnetostatic field, the anisotropy and the exchange field.

(4.108) can be derived by calculating the variation of the Gibbs free energy (3.8) and then setting it to zero with respect to the following boundary condition

$$|\boldsymbol{\mu}(\boldsymbol{r}, t)| = V_\mathrm{c} M_\mathrm{s} \quad . \tag{4.109}$$

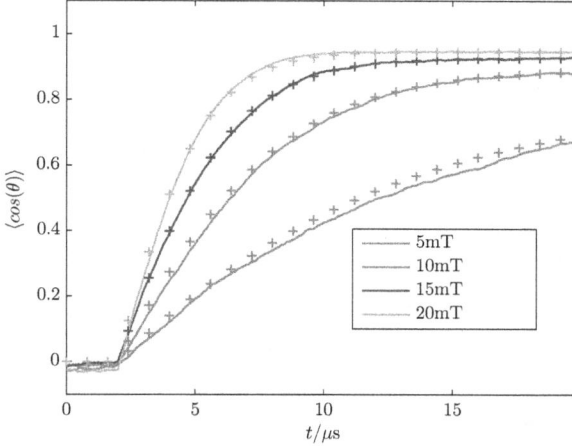

Fig. 4.2: Comparison between the results obtained by solving the Fokker Planck equation (+) or by directly integrating the Langevin equation (−). Here only 1000 particles have been simulated, so that a difference is visible. At the time $t = 2\,\mu s$ different magnetic fields are switched and the used parameters are: $T = 300\,\text{K}$, $d_{\text{c}} = 25\,\text{nm}$ and $d_{\text{h}} = 40\,\text{nm}$.

This boundary condition is justified by the domination of the exchange energy well below the Curie temperature.

Landau and Lifshitz supposed the time evolution of the magnetization to be proportional to the deviation out of the thermodynamic equilibrium

$$\frac{d}{dt}\boldsymbol{\mu} = -\gamma\boldsymbol{\mu} \times \boldsymbol{H}^{\text{eff}} \quad , \tag{4.110}$$

whereas γ is the product of the gyromagnetic ratio of an electron γ_e and the vacuum permeability μ_0

$$\gamma = \gamma_e \cdot \mu_0 = 1.76 \cdot 10^{11}\frac{\text{As}}{\text{kg}} \cdot 4\pi 10^{-7}\frac{\text{kg m}}{\text{As}^2} = 2.2 \cdot 10^5 \frac{\text{m}}{\text{As}} \, . \tag{4.111}$$

Since equation (4.110) only describes a precessional motion, a phenomenological damping term has to be added, which still satisfies the boundary condition (4.109)

$$\frac{\mathrm{d}}{\mathrm{d}t}\boldsymbol{\mu} = -\gamma\boldsymbol{\mu} \times \boldsymbol{H}^{\text{eff}} - \kappa\frac{\gamma}{M_s V_{\text{c}}}\boldsymbol{\mu} \times (\boldsymbol{\mu} \times \boldsymbol{H}^{\text{eff}}) \quad . \tag{4.112}$$

κ is a dimensionless damping constant. This equation exhibits unphysical behavior in the case of very large damping constants. Gilbert corrected this defect, which results

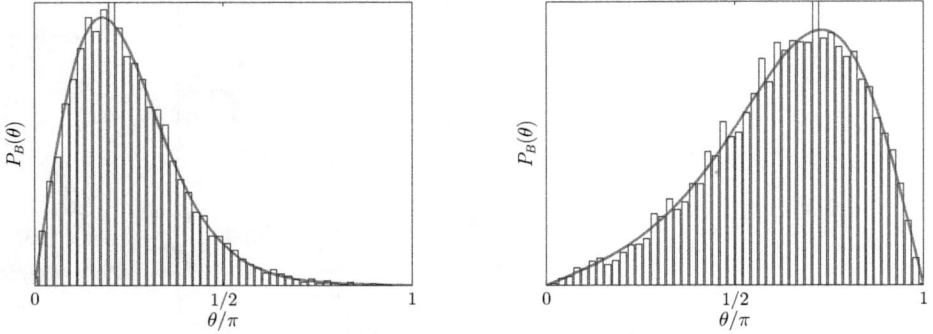

Fig. 4.3: Both figures show $P_{\mathrm{B}}(\theta)$ simulated with 10000 particles with $T = 310\,\mathrm{K}$, whereas the blue line is the result of the numerical evaluation of (4.107). The left figure is simulated by the Euler-Maruyama integration scheme with particles, specified by $d_{\mathrm{c}} = 15\,\mathrm{nm}$, $d_{\mathrm{h}} = 25\,\mathrm{nm}$ and $H_z = 15\,\mathrm{mT}/\mu_0$. The right figure visualizes the result obtained by the Heun integration scheme of particles with $d_{\mathrm{c}} = 10\,\mathrm{nm}$, $d_{\mathrm{h}} = 20\,\mathrm{nm}$ and $H_z = -20\,\mathrm{mT}/\mu_0$.

in the following equation, known as the Landau-Lifshitz-Gilbert Equation.

$$\frac{\mathrm{d}}{\mathrm{d}t}\boldsymbol{\mu} = \underbrace{-\frac{\gamma}{1+\alpha^2}\boldsymbol{\mu} \times \boldsymbol{H}^{\mathrm{eff}}}_{\text{Precession}} - \underbrace{\frac{\alpha\gamma}{(1+\alpha^2)M_{\mathrm{s}}V_{\mathrm{c}}}\boldsymbol{\mu} \times (\boldsymbol{\mu} \times \boldsymbol{H}^{\mathrm{eff}})}_{\text{Damping}} \quad . \tag{4.113}$$

α is again a dimensionless damping constant. In the case of magnetite this damping constant is of about $\alpha \approx 0.1$ [18, 17]. If one considers only a single domain nanoparticle, effects due to the exchange energy vanish and the effective field term consists only of the following components

$$\boldsymbol{H}^{\mathrm{eff}} = \boldsymbol{H}^{\mathrm{ext}} + \boldsymbol{H}^{\mathrm{An}} + \boldsymbol{H}^{\mathrm{MS}} \quad . \tag{4.114}$$

The crystal anisotropy and magnetostatic energy are given by

$$\begin{aligned}
\boldsymbol{H}^{\mathrm{An}} + \boldsymbol{H}^{\mathrm{MS}} &= \frac{2K}{\mu_0 M_{\mathrm{s}}^2 V_{\mathrm{c}}}\boldsymbol{n}_x \cdot (\boldsymbol{n}_x \cdot \boldsymbol{\mu}) - \begin{pmatrix} \mathcal{N}_x & 0 & 0 \\ 0 & \mathcal{N}_y & 0 \\ 0 & 0 & \mathcal{N}_z \end{pmatrix} \cdot \frac{\boldsymbol{\mu}}{V_{\mathrm{c}}} \\
&= -\begin{pmatrix} \mathcal{N}_x - \frac{2K}{\mu_0 M_{\mathrm{s}}^2} & 0 & 0 \\ 0 & \mathcal{N}_y & 0 \\ 0 & 0 & \mathcal{N}_z \end{pmatrix} \cdot \frac{\boldsymbol{\mu}}{V_{\mathrm{c}}} \quad .
\end{aligned} \tag{4.115}$$

Here it has been assumed, that the symmetry axis of the crystal anisotropy coincides with one of the principle axes of the ellipsoid.

Restricting oneself to rotational symmetric ellipsoidal particles, this relation may be simplified to

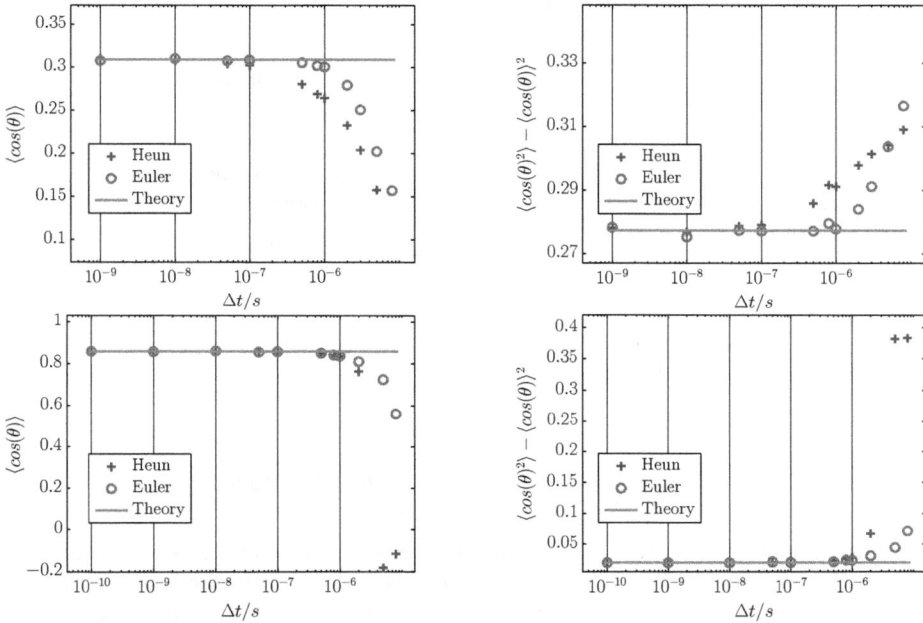

Fig. 4.4: Results of the evaluation of the time-step dependence of the Euler and the Heun Integration schemes. The figures visualize the deviation of the numerical results to the first two statistical moments of $P_B(\theta)$. The upper figures show the simulation of an equilibrium ensemble of 100000 Brownian particles with $d_c = 15\,\mathrm{nm}, d_h = 25\,\mathrm{nm}$, $T = 310\,\mathrm{K}$ and $H = 5\,\mathrm{mT}/\mu_0$. The figures below visualize the deviation of the mean magnetization and its variance, simulated with 10000 Brownian particles, whereas the parameters have been $d_c = 20\,\mathrm{nm}, d_h = 30\,\mathrm{nm}$, $T = 310\,\mathrm{K}$ and $H = 15\,\mathrm{mT}/\mu_0$.

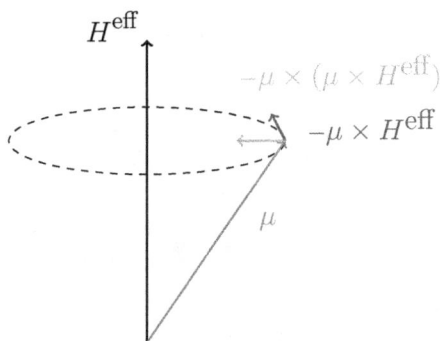

Fig. 4.5: The damped gyromagnetic rotation is initiated by the precessional term $-\boldsymbol{\mu} \times \boldsymbol{H}^{\text{eff}}$ and the damping term $-\boldsymbol{\mu} \times (\boldsymbol{\mu} \times \boldsymbol{H}^{\text{eff}})$. Both do not change the norm of the magnetic moment vector $|\boldsymbol{\mu}| = const.$

$$
\boldsymbol{H}^{\text{An}} + \boldsymbol{H}^{\text{MS}} = \begin{pmatrix} \mathcal{N}_{\|} - \frac{2K}{\mu_0 M_s^2} & 0 & 0 \\ 0 & \mathcal{N}_{\perp} & 0 \\ 0 & 0 & \mathcal{N}_{\perp} \end{pmatrix} \cdot \frac{\boldsymbol{\mu}}{V_c}
$$

$$
= \frac{1}{V_c} \left(((\mathcal{N}_{\|} - \frac{2K}{\mu_0 M_s^2}) - \mathcal{N}_{\perp})(\boldsymbol{n}\boldsymbol{\mu})\boldsymbol{n} + \mathcal{N}_{\perp}\boldsymbol{\mu} \right) \quad , \tag{4.116}
$$

where $\boldsymbol{n}$ is a unit vector, which points in the direction of the easy axis of the ellipsoid. The last term $\mathcal{N}_{\perp}\boldsymbol{\mu}$ has no influence on the motion of the magnetic moments and therefore one can justify the introduction of an effective anisotropy constant K_{eff}.

$$
\boldsymbol{H}^{\text{An}} + \boldsymbol{H}^{\text{MS}} = \frac{2K_{\text{eff}}}{\mu_0 M_s^2 V_c} \boldsymbol{n} \cdot (\boldsymbol{n} \cdot \boldsymbol{\mu}) \tag{4.117}
$$

Unfortunately, there is no general rule, which gives a hint how to include thermal fluctuations into (4.113). One may add it solely to the precessional term [42], add it to the damping and precessional term with different intensities or add it to the precessional and damping term with the same intensity. However it can be shown, that each possibility results in the same Fokker-Planck equation and therefore each possibility is physically equivalent [9].

In this thesis the thermal fluctuation is added to the precessional and damping term. In addition the reduced magnetic moment $\boldsymbol{m} = \boldsymbol{\mu}/|\boldsymbol{\mu}|$ is introduced. This finally gives

$$
\frac{\mathrm{d}}{\mathrm{d}t}\boldsymbol{m} = -\frac{\gamma}{1 + \alpha^2} \boldsymbol{m} \times (\boldsymbol{H}^{\text{eff}} + \boldsymbol{H}^{\text{th}}) - \frac{\alpha\gamma}{(1 + \alpha^2)} \boldsymbol{m} \times (\boldsymbol{m} \times (\boldsymbol{H}^{\text{eff}} + \boldsymbol{H}^{\text{th}})) \quad , \tag{4.118}
$$

with

$$\langle \boldsymbol{H}^{\text{th}}(t) \rangle = 0 \tag{4.119}$$
$$\langle H_i^{\text{th}}(t + \tau) H_j^{\text{th}}(t) \rangle = 2\mathcal{D}_N \delta_{ij} \delta(\tau) \quad . \tag{4.120}$$

Likewise to the Brownian diffusion the norm of $\boldsymbol{m}$ is only conserved, if the Stratonovich calculus is applied.

4.3.1 Fokker Planck Equation

By separating the deterministic and the stochastic part of (4.118)

$$\frac{\mathrm{d}}{\mathrm{d}t}\boldsymbol{m} = -\frac{\gamma}{1+\alpha^2}\boldsymbol{m} \times \boldsymbol{H}^{\text{eff}} - \frac{\alpha\gamma}{(1+\alpha^2)}\boldsymbol{m} \times (\boldsymbol{m} \times \boldsymbol{H}^{\text{eff}})$$
$$-\frac{\gamma}{1+\alpha^2}\boldsymbol{m} \times \boldsymbol{H}^{\text{th}} - \frac{\alpha\gamma}{(1+\alpha^2)}\boldsymbol{m} \times (\boldsymbol{m} \times \boldsymbol{H}^{\text{th}}) \quad , \tag{4.121}$$

one can identify the functions h_i and g_{ik} of the general SDE (4.37) [20]

$$h_i = \left[-\frac{\gamma}{1+\alpha^2}\boldsymbol{m} \times \boldsymbol{H}^{\text{eff}} - \frac{\alpha\gamma}{(1+\alpha^2)}\boldsymbol{m} \times (\boldsymbol{m} \times \boldsymbol{H}^{\text{eff}}) \right]_i \tag{4.122}$$

$$g_{ik} = -\frac{\gamma}{1+\alpha^2}\sum_{p=1}^{3}\epsilon_{ipk}m_p - \frac{\alpha\gamma}{(1+\alpha^2)}\sum_{n=1}^{3}\epsilon_{iln}m_l\epsilon_{nqk}m_q$$
$$= -\frac{\gamma}{1+\alpha^2}\sum_{p=1}^{3}\epsilon_{ipk}m_p - \frac{\alpha\gamma}{(1+\alpha^2)}(m_i m_k - \delta_{ik}\boldsymbol{m}^2) \quad . \tag{4.123}$$

By calculating

$$\sum_j \frac{\partial g_{jk}}{\partial m_j} = -\sum_j \frac{\gamma}{1+\alpha^2}\epsilon_{jjk} + \frac{\alpha\gamma}{(1+\alpha^2)}(\delta_{jj}m_k + \delta_{jk}m_i - \delta_{jk}2m_j)$$
$$= -\frac{\alpha\gamma}{(1+\alpha^2)}(3m_k + m_k - 2m_k) = -2\frac{\alpha\gamma}{(1+\alpha^2)}m_k \tag{4.124}$$

$$\sum_k g_{ik}\frac{\partial g_{jk}}{\partial m_j} = -\sum_k \left[\frac{\gamma}{1+\alpha^2}\epsilon_{ipk}m_p - \frac{\alpha\gamma}{(1+\alpha^2)}(m_i m_k - \delta_{ik}\boldsymbol{m}^2) \right] 2\frac{\alpha\gamma}{(1+\alpha^2)}m_k$$
$$= \frac{2\alpha\gamma^2}{(1+\alpha^2)^2}\left[\epsilon_{ipk} + \epsilon_{ikp}\right]m_k m_p + 2\left(\frac{\alpha\gamma}{(1+\alpha^2)}\right)^2\sum_k \left[m_i m_k^2 - \delta_{ik}\boldsymbol{m}^2 m_i\right]$$
$$= 2\left(\frac{\alpha\gamma}{(1+\alpha^2)}\right)^2\left[\boldsymbol{m}^2 m_i - \boldsymbol{m}^2 m_i\right] = 0 \tag{4.125}$$

and (with $a = \gamma/(1+\alpha^2)$, $b = \alpha\gamma/(1+\alpha^2)$ and $\boldsymbol{m}^2 = 1$)

$$
\begin{aligned}
\sum_{jk} g_{ik}g_{jk}\frac{\partial P}{\partial m_j} &= \sum_{jk}\left(a\epsilon_{ipk}m_p + b\left(m_im_k - \delta_{ik}\boldsymbol{m}^2\right)\right)\\
&\quad \cdot\left(a\epsilon_{jqk}m_q + b\left(m_jm_k - \delta_{jk}\boldsymbol{m}^2\right)\right)\frac{\partial P}{\partial m_j}\\
&= \sum_{jk}\Big[a^2\epsilon_{ipk}\epsilon_{jqk}m_pm_q +\\
&\quad + b^2\left(m_jm_k - \delta_{jk}\boldsymbol{m}^2\right)\left(m_im_k - \delta_{ik}\boldsymbol{m}^2\right) +\\
&\quad + ab\left(\epsilon_{ipk}m_p\left(m_jm_k - \delta_{jk}\boldsymbol{m}^2\right) + \epsilon_{jqk}m_q\left(m_im_k - \delta_{ik}\boldsymbol{m}^2\right)\right)\Big]\frac{\partial P}{\partial m_j}\\
&= \sum_{jk}\Big[-a^2\epsilon_{ipk}\epsilon_{kqj}m_pm_q +\\
&\quad + b^2\left(m_k^2m_im_j - \delta_{jk}\boldsymbol{m}^2m_im_k - \delta_{ik}\boldsymbol{m}^2m_jm_k + \delta_{ik}\delta_{jk}\boldsymbol{m}^2\boldsymbol{m}^2\right) +\\
&\quad + ab\left(-\epsilon_{ipk}m_p\delta_{jk}\boldsymbol{m}^2 - \epsilon_{jqk}m_q\delta_{ik}\boldsymbol{m}^2 + \epsilon_{ipk}m_pm_jm_k + \epsilon_{jqk}m_qm_im_k\right)\Big]\frac{\partial P}{\partial m_j}\\
&= \sum_{jk}\Big[-a^2(m_im_j - \delta_{ij}\boldsymbol{m}^2) + b^2\boldsymbol{m}^2(\delta_{ij}\boldsymbol{m}^2 - m_im_j)\Big]\frac{\partial P}{\partial m_j}\\
&\quad + \sum_{j}ab\Big[\left(-\epsilon_{ipj}m_p\boldsymbol{m}^2 - \epsilon_{jqi}m_q\boldsymbol{m}^2\right)\Big]\frac{\partial P}{\partial m_j}\\
&= \sum_{jk}\Big[-(1+\alpha^2)\left(\frac{\gamma}{1+\alpha^2}\right)^2(m_im_j - \delta_{ij}\boldsymbol{m}^2)\Big]\frac{\partial P}{\partial m_j}\\
&= -\left(\frac{\gamma^2}{1+\alpha^2}\right)\Big[\boldsymbol{m}\times\left(\boldsymbol{m}\times\frac{\partial}{\partial\boldsymbol{m}}\right)\Big]_i P
\end{aligned}
$$

$$(4.126)$$

one finds, that the Fokker-Planck Equation is given by:

$$
\begin{aligned}
\frac{\partial P(\boldsymbol{m},t)}{\partial t} &= \frac{\partial}{\partial\boldsymbol{m}}\Big[\frac{\gamma}{1+\alpha^2}\boldsymbol{m}\times\boldsymbol{H}^{\mathrm{eff}} + \frac{\alpha\gamma}{(1+\alpha^2)}\boldsymbol{m}\times(\boldsymbol{m}\times\boldsymbol{H}^{\mathrm{eff}}) -\\
&\quad -\left(\frac{\mathcal{D}_{\mathrm{N}}\gamma^2}{1+\alpha^2}\right)\left(\boldsymbol{m}\times\left(\boldsymbol{m}\times\frac{\partial}{\partial\boldsymbol{m}}\right)\right)P(\boldsymbol{m},t)\Big]
\end{aligned}
$$

$$(4.127)$$

Again the stationary solution of P

$$\frac{\partial P(t)}{\partial t} = 0 \qquad (4.128)$$

has to match the Boltzmann distribution

$$P_0 \propto \exp\left(-\frac{\mathcal{H}(\boldsymbol{m})}{k_{\mathrm{B}}T}\right) \qquad . \qquad (4.129)$$

Taking into account the fact that the effective field is given by the derivative of the Hamiltonian $\mathcal{H}(\boldsymbol{m})$ [20]

$$\boldsymbol{H}^{\text{eff}} = -\frac{1}{\mu_0|\boldsymbol{\mu}|}\frac{\partial\mathcal{H}(\boldsymbol{\mu})}{\partial\boldsymbol{m}}, \tag{4.130}$$

one findes that the derivative of the equilibrium distribution is given by

$$\frac{\partial P_0}{\partial\boldsymbol{m}} = \frac{\mu_0|\boldsymbol{\mu}|}{k_{\text{B}}T}\boldsymbol{H}^{\text{eff}}P_0 \quad . \tag{4.131}$$

The first term of the Fokker Planck equation (4.127) vanishes

$$\begin{aligned}
\frac{\partial}{\partial\boldsymbol{m}}\left[\frac{\gamma}{1+\alpha^2}\boldsymbol{m}\times\boldsymbol{H}^{\text{eff}}\right] &= \frac{\gamma}{1+\alpha^2}\sum_i\frac{\partial}{\partial m_i}\epsilon_{ilj}m_lH_jP_0 \\
&= \frac{\gamma}{1+\alpha^2}\sum_i\left(\epsilon_{iij}H_jP_0 + \epsilon_{ilj}m_lH_j\frac{\partial P_0}{\partial m_i}\right) \\
&= \frac{\gamma}{1+\alpha^2}\sum_i\frac{m_0}{k_{\text{B}}T}\epsilon_{ilj}m_lH_iH_jP_0 \\
&= 0
\end{aligned} \tag{4.132}$$

and therefore the equilibrium condition is given by:

$$0 = \left[\frac{\alpha\gamma}{(1+\alpha^2)}\boldsymbol{m}\times(\boldsymbol{m}\times\boldsymbol{H}^{\text{eff}})P_0(\boldsymbol{m}) - \left(\frac{\mu_0|\boldsymbol{\mu}|\mathcal{D}_{\text{N}}\gamma^2}{k_{\text{B}}T(1+\alpha^2)}\right)\left(\boldsymbol{m}\times\left(\boldsymbol{m}\times\boldsymbol{H}^{\text{eff}}\right)\right)P_0(\boldsymbol{m})\right]$$

In general this equation is only valid if

$$\mathcal{D}_{\text{N}} = \frac{k_{\text{B}}T\alpha}{\mu_0|\boldsymbol{\mu}|\gamma} \quad . \tag{4.133}$$

It is worth noting, that the final expression of the diffusion constant does not depend on a specific Hamiltonian and therefore the diffusion constant $\mathcal{D}_{\text{N}}$ would remain the same even if inter-particle interactions would be included.

4.3.2 Numerical Integration

Euler-Maruyama

First, it is necessary to calculate the spurious drift term with (4.123), $a = \gamma/(1+\alpha^2)$, $b = \alpha\gamma/\left((1+\alpha^2)M_sV_c\right)$ and $\boldsymbol{m}^2 = 1$, one gets:

$$\begin{aligned}
\sum_{j=1}^3\sum_{i=1}^3 g_{ji}\frac{\partial g_{ki}}{\partial m_j} &= \sum_{j=1}^3\sum_{i=1}^3\left(-a\sum_p^3\epsilon_{jpi}m_p - b(m_jm_i - \delta_{ij}\boldsymbol{m}^2)\right) \\
&\quad (-a\epsilon_{ijk} - b(\delta_{ij}m_k + \delta_{jk}m_i - 2\delta_{ik}m_j))
\end{aligned} \tag{4.134}$$

In order to maintain a better overview this should be calculated by parts

$$\sum_{j=1}^{3}\sum_{i=1}^{3} a^2 \sum_{p}^{3} \epsilon_{ijk}\epsilon_{jpi}m_p = -2a^2 \sum_{p}^{3} \delta_{pk}m_p = -2a^2 m_k$$

$$\sum_{j=1}^{3}\sum_{i=1}^{3} b^2 m_j m_i (\delta_{ij}m_k + \delta_{jk}m_i - 2\delta_{ik}m_j) = b^2 \left(m_k \boldsymbol{m}^2 + m_k \boldsymbol{m}^2 - 2m_k \boldsymbol{m}^2 \right) = 0$$

$$\sum_{j=1}^{3}\sum_{i=1}^{3} b^2 \delta_{ji} \boldsymbol{m}^2 (\delta_{ij}m_k + \delta_{jk}m_i - 2\delta_{ik}m_j) = b^2 \left(-3m_k \boldsymbol{m}^2 - m_k \boldsymbol{m}^2 + 2m_k \boldsymbol{m}^2 \right)$$

$$= -2b^2 \boldsymbol{m}^2 m_k$$

$$\sum_{j=1}^{3}\sum_{i=1}^{3} ab \sum_{p}^{3} \epsilon_{jpi}m_p (\delta_{ij}m_k + \delta_{jk}m_i - 2\delta_{ik}m_j) = 0$$

$$\sum_{j=1}^{3}\sum_{i=1}^{3} ab\epsilon_{ijk}(m_j m_i - \delta_{ij}\boldsymbol{m}^2) = 0 \quad .$$

$$(4.135)$$

The result

$$\sum_{j=1}^{3}\sum_{i=1}^{3} g_{ji}\frac{\partial g_{ki}}{\partial m_j} = -2\gamma^2 \frac{1}{(1+\alpha^2)} m_k \qquad (4.136)$$

then leads to the Euler-Maruyama integration scheme of the Néel diffusion

$$\boldsymbol{m}(t+\Delta t) = \boldsymbol{m}(t)$$
$$- \frac{\gamma}{1+\alpha^2}\left[\boldsymbol{m}(t) \times \boldsymbol{H}^{\mathrm{eff}}(t) + \alpha\boldsymbol{m}(t) \times (\boldsymbol{m}(t) \times \boldsymbol{H}^{\mathrm{eff}}(t)) - 2\mathcal{D}_{\mathrm{N}}\gamma\boldsymbol{m}(t)\right]\Delta t$$
$$- \frac{\gamma\sqrt{2\mathcal{D}_{\mathrm{N}}}}{1+\alpha^2}\left[\boldsymbol{m}(t) \times \Delta\boldsymbol{W} + \alpha\boldsymbol{m}(t) \times (\boldsymbol{m}(t) \times \Delta\boldsymbol{W})\right] \quad .$$

$$(4.137)$$

Heun

The predictor step is given by

$$\overline{\boldsymbol{m}}(t+\Delta t) = \boldsymbol{m}(t)$$
$$- \frac{\gamma}{1+\alpha^2}\left[\boldsymbol{m}(t) \times \boldsymbol{H}^{\mathrm{eff}}(t) + \alpha\boldsymbol{m}(t) \times (\boldsymbol{m}(t) \times \boldsymbol{H}^{\mathrm{eff}}(t))\right]\Delta t$$
$$- \frac{\gamma\sqrt{2\mathcal{D}_{\mathrm{N}}}}{1+\alpha^2}\left[\boldsymbol{m}(t) \times \Delta\boldsymbol{W} + \alpha\boldsymbol{m}(t) \times (\boldsymbol{m}(t) \times \Delta\boldsymbol{W))}\right]$$

$$(4.138)$$

and the corrector reads

$$\boldsymbol{m}(t + \Delta t) = \boldsymbol{m}(t)$$

$$- \frac{\gamma}{2(1 + \alpha^2)} \Bigg[\boldsymbol{m}(t) \times \boldsymbol{H}^{\text{eff}}(t) + \alpha \boldsymbol{m}(t) \times (\boldsymbol{m}(t) \times \boldsymbol{H}^{\text{eff}}(t))$$

$$+ \overline{\boldsymbol{m}}(t + \Delta t) \times \boldsymbol{H}^{\text{eff}}(t + \Delta t)$$

$$+ \alpha \overline{\boldsymbol{m}}(t + \Delta t) \times (\overline{\boldsymbol{m}}(t + \Delta t) \times \boldsymbol{H}^{\text{eff}}(t + \Delta t)) \Bigg] \Delta t$$

$$- \frac{\gamma \sqrt{2\mathcal{D}_{\text{N}}}}{2(1 + \alpha^2)} \Bigg[\boldsymbol{m}(t) \times \Delta \boldsymbol{W} + \alpha \boldsymbol{m}(t) \times (\boldsymbol{m}(t) \times \Delta \boldsymbol{W})$$

$$+ \overline{\boldsymbol{m}}(t + \Delta t) \times \Delta \boldsymbol{W} + \alpha \overline{\boldsymbol{m}}(t + \Delta t) \times (\overline{\boldsymbol{m}}(t + \Delta t) \times \Delta \boldsymbol{W}) \Bigg] .$$

$$(4.139)$$

4.3.3 Tests

Analogue to the Brownian diffusion, the integration schemes of the Néel diffusion can only be judged according to the weak order convergence criterion.

A good test system is an equilibrium particle ensemble of particles with an uniaxial anisotropy, whereas the angle χ_{f} between the easy axis and the magnetic field is fixed. The probability distribution of the angle θ between the magnetic moment and the applied external magnetic field should coincide with

$$P_{\text{N}}(\theta) = \frac{1}{Z} \int_0^{2\pi} \int_0^{2\pi} e^{(\lambda(\cos(\chi_{\text{f}})\cos(\theta) + \sin(\chi_{\text{f}})\sin(\theta)\cos(\alpha - \beta))^2 + \rho\cos(\theta))} \sin(\chi_{\text{f}}) \sin(\theta) \mathrm{d}\beta \mathrm{d}\alpha \quad .$$

$$(4.140)$$

This can be derived from (3.41). It was found that both integration schemes match very well with the theoretical predicted probability distribution. Some results are presented in Figure (4.6). Once again, the time step dependence can be checked by evaluating the deviation of the simulated mean value and variance of the particle ensemble. The Heun solver obviously has better properties than the Euler-Maruyama solver. The different result compared to the Brownian diffusion may originate in the precessional character of the Landau Lifshitz Gilbert equation. The precessional part of the motion of the magnetic moments can be described in the absence of thermal agitation by the Lamor frequency, which is given by

$$\omega_L = \gamma H_{\text{eff}} \quad . \tag{4.141}$$

Therefore, the spiraling towards the energy minimum is characterized by the following time

$$\tau_{\text{N}} = 1/\alpha \omega_L \quad . \tag{4.142}$$

Garcias-Palacios et al [20] proposed to choose the integration time step according to

$$\Delta t \approx 0.01 \tau_{\text{N}} \quad , \tag{4.143}$$

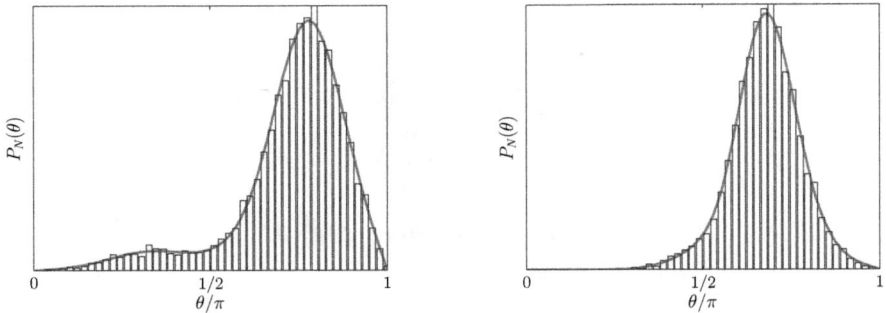

Fig. 4.6: Both figures visualizes $P_N(\theta)$ simulated with 10000 particles at a temperature of $T = 310\,\text{K}$, where the blue line is the result of the numerical integration of (4.140). The left figure is generated with the Heun integration scheme of particles with $d_c = 15\,\text{nm}$, $K = 10^4\,\text{J/m}^3$, $H_z = -10\,\text{mT}/\mu_0$ and $\chi_f = 1/4\pi$. The right one visualizes the result obtained with the Heun integration scheme of particles with $d_c = 20\,\text{nm}$, $K = 7.8 \cdot 10^3\,\text{J/m}^3$, $H_z = -10\,\text{mT}/\mu_0$ and $\chi_f = 2/5\pi$.

which has been found to be correct and therefore this rule is subsequently used to choose the right size for the time step.

4.4 Combined Rotation

The relaxation processes present in a ferrofluid require a model, which incorporates both, the Néel and the Brownian diffusion [33, 5]. The procedure, as presente for the Néel and Brownian diffusion - formulate a Langevin equation - derive the Fokker-Planck equation - finally calculate the diffusion constant, can not be transferred to the problem of the combined rotation. This is due to the connection of the Brownian and the Néel diffusion by the magnetic field. Both processes can only be separated in the case of very weak magnetic fields [75]. Thus, the combined rotation has to be simulated by a numerical integration scheme, which integrates simultaneously the Brownian and the Néel diffusion equation. Here it is important that each integration step of the mechanical rotation affects the integration step of the Néel diffusion and vice versa.

The Brownian diffusion leads to a change of the orientation of the particle itself and therefore also changes the direction of the easy axis n. This reorientation of n modulates the solution of the Néel diffusion, which results in a different orientation of the particles magnetic moment to the easy axis. The changed orientation of the magnetic moment has of course again influence on Brownian rotation.

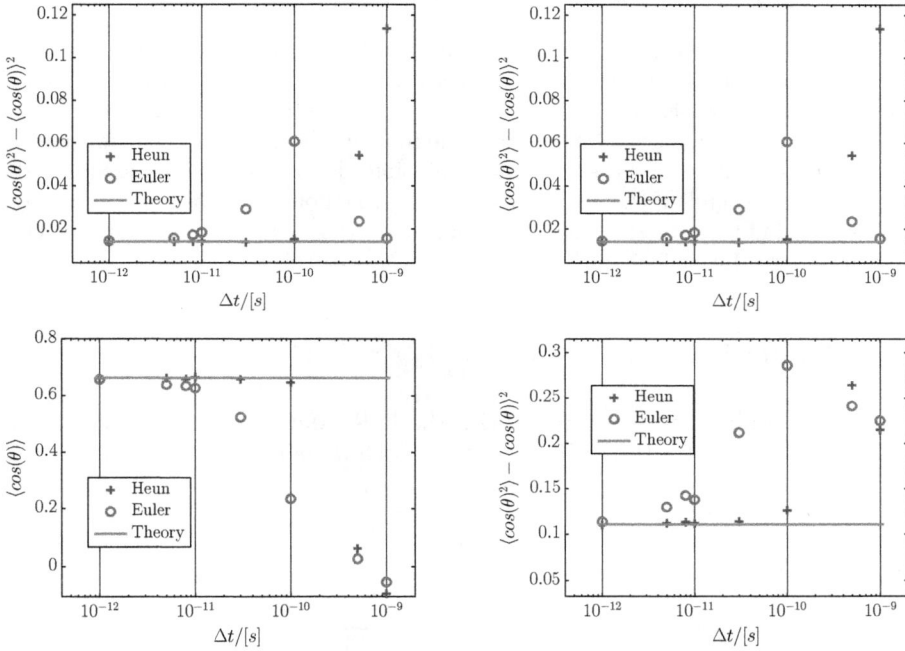

Fig. 4.7: Results of the evaluation of the time step dependence of the Euler and the Heun Integration schemes. The upper figures show the simulation of an equilibrium ensemble of 100000 particles with $d_c = 20\,\mathrm{nm}$, $K = 0.7 \cdot 10^4\mathrm{J/m}^3$, $\theta_n = 1/5\pi$, $T = 310\,\mathrm{K}$ and $H = -15\,\mathrm{mT}/\mu_0$. The figures below visualize the deviation of the mean magnetization and its variance, simulated with 10000 particles, where the parameters have been $d_c = 20\,\mathrm{nm}$, $K = 1.2 \cdot 10^4\mathrm{J/m}^3$, $\theta_n = 1/5\pi$, $T = 310\,\mathrm{K}$ and $H = 5\,\mathrm{mT}/\mu_0$.

The equation of motion of the easy axis n can be calculated by the equation

$$\frac{\mathrm{d}}{\mathrm{d}t}n = \frac{|\boldsymbol{\mu}|}{\zeta}\left[(m \times H^{\mathrm{ext}}) \times n\right] - \frac{1}{\zeta}\left[n \times \boldsymbol{\lambda}\right] \quad . \tag{4.144}$$

This can be derived by using

$$\frac{\mathrm{d}}{\mathrm{d}t}n = \boldsymbol{\omega} \times n \tag{4.145}$$

instead of (4.62).

The model of combined rotation should be able to handle the limits of a fixed particle or a blocked magnetic moment. If the particles are fixed, for example because of a large viscosity, the dynamic is solely affected by the the Néel diffusion and if the the particles magnetic moment is blocked, for example due to a large anisotropy, the magnetization dynamic is only driven by the Brownian diffusion. Furthermore, it is assumed that both motions decouple if the magnetic field is switched off. This assumption of course neglects the influence of the magnetic moment on the mechanical torque, known as the Einstein-de Haas effect.

4.4.1 Numerical Integration

The Brownian diffusion is integrated with the Euler-Maruyama solver and the Néel diffusion is integrated with the Heun scheme. Each time integration step Δt consists of the following steps.

1. Calcualtion of the next time step values:
   Brownian Rotation:

$$\begin{aligned} n(\tau) &\to n(\tau + \Delta t) \\ m(\tau) &\to m_{\mathrm{B}}(\tau + \Delta t) \end{aligned} \tag{4.146}$$

   Néel Rotation:

$$m(\tau) \to m_{\mathrm{N}}(\tau + \Delta t) \tag{4.147}$$

2. Combination of both magnetic moments:

$$m(\tau + \Delta t) = (m_{\mathrm{B}}(\tau + \Delta t) + m_{\mathrm{N}}(\tau + \Delta t)) - m(\tau) \tag{4.148}$$

3. Projection step:

$$\begin{aligned} m(\tau + \Delta t) &= \frac{m(\tau + \Delta t)}{|m(\tau + \Delta t)|} \\ n(\tau + \Delta t) &= \frac{n(\tau + \Delta t)}{|n(\tau + \Delta t)|} \end{aligned} \tag{4.149}$$

Here the subscripts B and N stand for Brownian and Néel rotation.

Due to the second step, the integration scheme reduces naturally to the Euler-Maruyama integration scheme of the Brownian diffusion if the magnetic moments are blocked. Likewise this it is equal the Heun integration scheme of the Néel diffusion if the particles are fixed. The last step ensures the conservation of the norm of the magnetic norm. Furthermore it has been found, that this step enhances numerical stability. Nevertheless it has also been tested, that this projection step is actually not needed to conserve the length of the magnetic vector, since the Stratonovich interpretation is applied.

4.4.2 Tests

Considering an equilibrium ensemble of ferrofluid-particles with uniaxial anisotropy the angular distribution of the magnetic moments has to coincide with the following probability distribution (3.4)

$$P_C^m(\theta) = \frac{1}{Z} \sin(\theta) \exp\left(\frac{\mu_0 V_c M_s H_z \cos(\theta)}{k_B T}\right) \quad . \tag{4.150}$$

An even more reliable test may be constructed by comparing the probability distribution of the angle χ between the easy axis and the magnetic field with the theoretical predicted probability distribution P_C^n:

$$P_C^n(\chi) = \frac{1}{Z} \int_0^\pi \int_0^{2\pi} \int_0^{2\pi} e^{(\lambda(\cos(\chi)\cos(\theta) + \sin(\chi_f)\sin(\theta)\cos(\alpha-\beta))^2 + \rho\cos(\theta))} \sin(\chi)\sin(\theta) d\beta d\alpha d\theta \tag{4.151}$$

This provides a deeper insight into the properties of the integration scheme, since the Néel and the Brownian diffusion are solely connected via the anisotropy term.

It has been found, that the simulation results coincide well with the theoretical predicted probability distribution (4.4.2).

Furthermore, it has been checked, that the limits of the blocked and the rigid particle are correctly simulated, see figure (4.9) and figure (4.10).

The time step size has been choosen according to the rule of Garcias-Palacios (4.143), because the Néel diffusion requires a smaller time step compared to the Brownian diffusion.

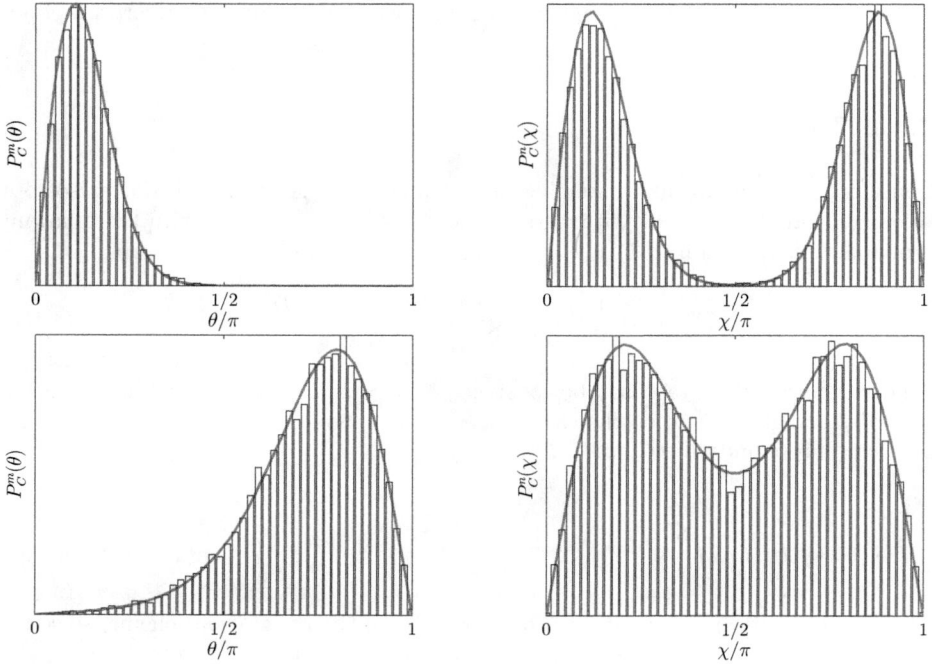

Fig. 4.8: Simulated equilibrium probability distribution of χ and θ. The results have been obtained by simulating 10000 particles at a temperature of $T = 310\,\text{K}$. Each blue line represents the numerical evaluation of P_C^n or P_C^m, respectively. The simulation parameter of the upper two figures have been: $K = 1 \cdot 10^4\,\text{J/m}^3$, $d_c = 25\,\text{nm}$, $H = 0.01\,\text{mT}/\mu_0$. The figures in the lower line have been simulated with $K = 4.9 \cdot 10^3\,\text{J/m}^3$, $d_c = 20\,\text{nm}$ and $H = -0.005\,\text{mT}/\mu_0$. It is apparent, that the distribution $P_\text{C}^\text{n}(\chi)$ is always symmetric with respect to $\chi = 1/2\pi$.

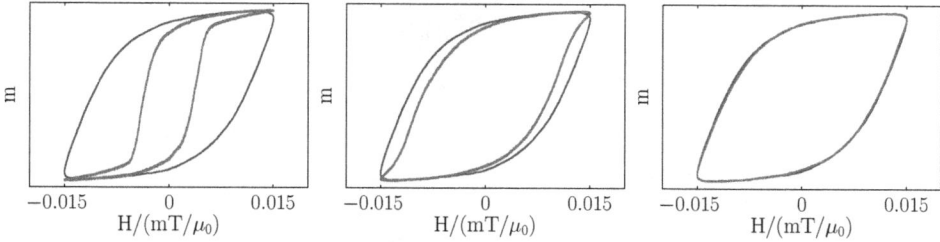

Fig. 4.9: Hysteresis loops of a $d_c = 25\,nm$ particle with a $d_h = 35\,nm$ shell. The blue curves are the solution of the Brownian diffusion. The green curves have been calculated with the combined model, whereby the anisotropy values have been (from left to right) $K_1 = 0.5 \cdot 10^4\,J/m^3$, $K_2 = 1 \cdot 10^4\,J/m^3$ and $K_3 = 10 \cdot 10^4 J/m^3$. If the particle is blocked due to a high anisotropy, both solution coincide. Additional simulation parameters, which have been used: $H = 0.015\,mT/\mu_0$, $f = 25\,kHz$, $\eta = 0.9 \cdot 10^{-3}\,Ns/m^2$, $T = 300\,K$, $N = 1000$.

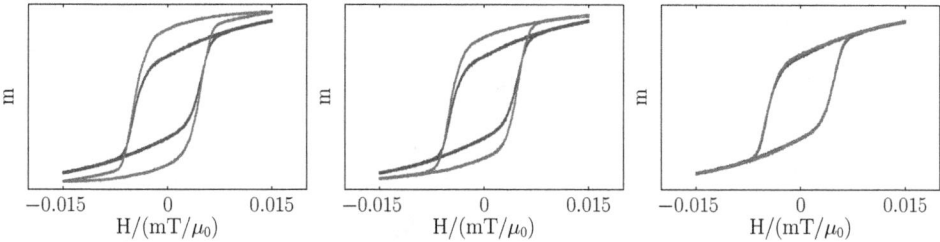

Fig. 4.10: Hysteresis loops of a $d_c = 25\,nm$ particle with an anisotropy constant of $K = 0.5 \cdot 10^4\,J/m^3$. The blue curves are the solutions of the Néel diffusion and the green curves have been calculated with the combined rotation model. The different hydrodynamical diameters have been $d_h = 30\,nm$, $d_h = 70\,nm$ and $d_h = 1\,\mu m$ (from left to right). Both solutions coincide if the particle can not rotate bodily due to a high friction. Additional simulation parameters, which have been used: $H = 0.015\,mT/\mu_0$, $f = 70\,kHz$, $\eta = 0.9 \cdot 10^{-3}\,Ns/m^2$, $T = 300\,K$, $N = 1000$.

5 Approximated Simple Particle Models

The particle functions obtained by the stochastic particle model are naturally noise corrupted and computationally expensiv. Therefore it is hardly an appropriate model to fit experimental data. Nevertheless it can be used to obtain a deep insight into the dynamic of MPI contrast agents.

There is need for a simple particle model, which in future can be used to generate the system function (2.3) or to enable fast characterization of the performance of a specific contrast agent. Such a model should be able to handle the nonlinear magnetization behavior combined with rate-dependent hysteresis phenomena. Furthermore, it should be possible to extrapolate the magnetization behavior from an MPS to any arbitrary MPI scanner.

5.1 Equilibrium Particle Model

The easiest way is to consider the system to be always in thermodynamic equilibrium. As presented above (see 3.4) such a system is then described correctly by the Langevin equation and therefore the magnetization is given by

$$\boldsymbol{M}(\boldsymbol{H}, T) = \frac{N|\boldsymbol{\mu}|}{V_{\mathrm{s}}} \mathcal{L}(\rho) \boldsymbol{e}_H \quad . \tag{5.1}$$

Since a common ferrofluid is not monodisperse, this equation may be extended to a polydisperse one.

$$\boldsymbol{M}(\boldsymbol{H}, T) = \frac{N}{V_{\mathrm{s}}} \int_0^\infty \mathcal{P}(d_{\mathrm{c}}) |\boldsymbol{\mu}(d_{\mathrm{c}})| \mathcal{L}(\rho) \boldsymbol{e}_H \mathrm{d}d_{\mathrm{c}} \tag{5.2}$$

$\mathcal{P}(d_{\mathrm{c}})$ is a normalized distribution function, which is often assumed to be a log-normal distribution. This presumption is often justified by the presence of sedimentation and agglomeration processes, which are of importance during the production process of a ferrofluid [35]. This particle model is in extensive use in MPI [38, 31, 7] but it totally excludes all hysteresis phenomena.

5.2 Extended Debye Model

The Debye model of relaxation can be derived by a linear approximation of the probability distribution (4.81)

$$P(\theta, t) = a_1 + a_2(t)\cos(\theta) \quad , \tag{5.3}$$

which is the solution of the dynamics described by the Fokker Planck equation of the Brownian rotation (4.80) [75]. If the magnetic field is given by

$$H(t) = H_0 e^{i\omega t} \quad , \tag{5.4}$$

one can deduce, that the mean magnetic moment is given by

$$|\boldsymbol{\mu}|\langle\cos(\theta)\rangle = \frac{|\boldsymbol{\mu}|^2 \mu_0 H_0 e^{i\omega t}}{3 k_{\mathrm{B}} T (1 + i\omega\tau_{\mathrm{B}})} \quad . \tag{5.5}$$

This formula is only valid within linear response theory and therefore it is not able to handle the nonlinear magnetization response. One may now think of replacing the linear susceptibility by the nonlinear Langevin function. This leads to the following equation

$$M(\rho) = (\chi' + i\chi'')\frac{|\boldsymbol{\mu}|N}{V_{\mathrm{s}}}\mathcal{L}(\rho) \quad , \tag{5.6}$$

where the susceptibility components χ' and χ'' are defined by:

$$\begin{aligned} \chi' &= 1/(1 + (\omega\tau_{\mathrm{B}})^2) \\ \chi'' &= \omega\tau_{\mathrm{B}}/(1 + (\omega\tau_{\mathrm{B}})^2) \end{aligned} \tag{5.7}$$

These equations have also been used to investigate the MPI signal [59, 56, 55]. They are often accompanied by the assumption of log-normal size distributed particles. The effective relaxation times can be calculated by (3.35).

5.3 ODE-based Particle Model

A different approach to approximate the magnetization behavior is to consider the magnetization to be slightly out of equilibrium and therefore expand the magnetization in a Taylor series

$$\boldsymbol{M}(t) = \frac{|\boldsymbol{\mu}|N}{V_{\mathrm{s}}}\mathcal{L}(\rho(t))\boldsymbol{e}_{\mathrm{H}} - \tau\frac{\mathrm{d}\boldsymbol{M}(t)}{\mathrm{d}t} + \dots \quad , \tag{5.8}$$

which leads to an ordinary differential equation (ODE).

$$\frac{\mathrm{d}\boldsymbol{M}(t)}{\mathrm{d}t} = \frac{1}{\tau}\left(\frac{|\boldsymbol{\mu}|N}{V_{\mathrm{s}}}\mathcal{L}(\rho(t))\boldsymbol{e}_{\mathrm{H}} - \boldsymbol{M}\right) \quad . \tag{5.9}$$

That equation has also been used by Shliomis to describe hydrodynamic effects of a ferrofluid [68].

There are reasonable physical limits of this equation. First, magnetization response coincides with the Langevin equation, if the relaxation time τ is small compared to a different characteristic time.

$$\lim_{\tau \to 0} \boldsymbol{M} = \frac{|\boldsymbol{\mu}|N}{V_\mathrm{s}} \mathcal{L}(\rho(t)) \boldsymbol{e}_\mathrm{H} \qquad (5.10)$$

Next, it coincides with the Debye theory in the limit of small magnetic fields. If there are only low magnetic field strength, the Langevin equation may be approximated by a linear susceptibility and therefore the ODE (5.9) simplifies to

$$\frac{\mathrm{d}\boldsymbol{M}(t)}{\mathrm{d}t} = \frac{1}{\tau} (\chi_0 \boldsymbol{H} - \boldsymbol{M}) \quad . \qquad (5.11)$$

Assuming that the magnetic field is given by (5.4) and additionally a Fourier transformation of (5.11) is performed, the result is

$$i\omega \tilde{M} = \frac{1}{\tau} \left(\chi_0 \tilde{H} - \tilde{M} \right) \qquad (5.12)$$

and therefore the linear susceptibility is given by

$$\chi(\omega) = \frac{\tilde{M}}{\tilde{H}} = \frac{1}{(1 + i\tau\omega)} \chi_0 \quad . \qquad (5.13)$$

Up to now, the relaxation time τ is assumed to be constant, though it is very likely that it depends on $\rho = \mu_0 |\boldsymbol{\mu}|/(k_\mathrm{B}T)$. The magnetization response to a magnetic field, which is switched on at a specific time, will clearly depend on its field strength. For example, one might compare two different field strengths $H_1 = 50\,\mathrm{mT}/\mu_0$ and $H_2 = 100\,\mathrm{mT}/\mu_0$. The equilibrium magnetization will coincide, since the ferrofluid will be saturated at both field strengths, but the equilibrium state will be reached earlier if the higher field strength is applied.

To investigate this effect, it is necessary to solve the ODE (5.9) in the case of a constant magnetic field, which is applied at the time $t = 0$. This result is given by

$$M(t) = \mathcal{L}(\rho) \left[1 - e^{-t/\tau} \right] \quad . \qquad (5.14)$$

The effective relaxation time τ_eff can then be estimated by fitting (5.14) to solutions, which are simulated by the stochastic particle model. Unfortunately, the combined rotation model cannot be taken, since it is a combination of two diffusion processes and therefore it is not possible to approximate its solution by (5.14). Furthermore, it is also not possible to investigate the Néel diffusion, since the thermodynamic equilibrium function of rigid particles does not coincide with the Langevin function. Thus, only the Brownian diffusion has been considered. This is of course a quite rough approximation, where hopefully the combined rotation or the Néel diffusion behaves equal to the Brownian diffusion.

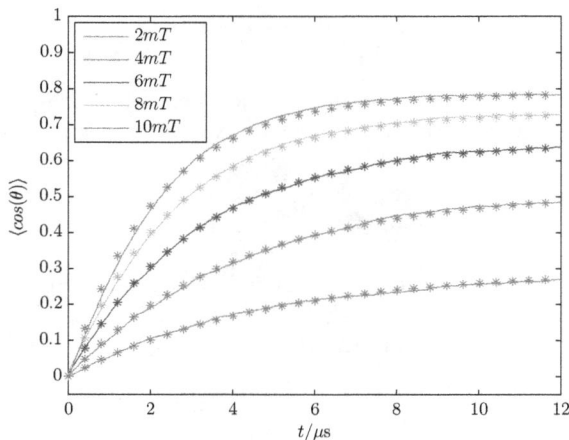

Fig. 5.1: Magnetization response of a Brownian particle with $d_c = 20\,\text{nm}$ and $d_h = 25\,\text{nm}$ to a magnetic field with different field strengths, applied at $t = 0$. The asterisks indicates the fit solution of (5.14), whereas τ has been the only free parameter. The deviation between both gets larger as the field strength increases.

A least square fit has been used to fit τ from (5.14) to the solution of the Brownian diffusion. Different field strength and core diameters have been simulated and with it different values of ρ. Some results are visualized in figure (5.1). It has been found that the equation (5.14) is a good approximation, if the change of ρ is small ($\rho < 3$). This is understandable since (5.9) is correct, only if the deviation out of the equilibrium is small. Additionally, different hydrodynamic diameters d_h have been simulated and thus different relaxation times, see figure (5.2). A linear dependency has been found, which can be nicely approximated by

$$\tau_{\text{eff}} = \tau_0(1 - 0.125 \cdot \rho) \quad . \tag{5.15}$$

This result coincides well with that one published by T. Yoshida [76].

Since it is to be expected that the contrast agent consists of different particles with different diameters and different relaxation times, the contrast agent should be assumed to be polydisperse. This can be acknowledged by weighting the obtained solutions with a normalized distribution function $\mathcal{P}(d_c, \tau_0)$

$$M(t) = \frac{N}{V_s} \int_0^\infty \int_0^\infty \mathcal{P}(d_c, \tau_0)\langle\boldsymbol{\mu}(\tau_0, d_c, t)\rangle \mathrm{d}d_c\mathrm{d}\tau_0 \quad , \tag{5.16}$$

whereas $\langle\boldsymbol{\mu}(\tau_0, d_c, t)\rangle$ is the solution of

$$\frac{\mathrm{d}\langle\boldsymbol{\mu}(\tau_0, d_c, t)\rangle}{\mathrm{d}t} = \frac{1}{\tau_0(1 - 0.125 \cdot \rho(t))}\left(|\boldsymbol{\mu}|\mathcal{L}(\rho(t))\boldsymbol{e}_{\text{H}}(t) - \langle\boldsymbol{\mu}(\tau_0, d_c, t)\rangle\right) \quad . \tag{5.17}$$

It is hardly possible to make any assumptions about $\mathcal{P}(d_c, \tau_0)$, but it is to be expected, that the distribution is a smooth function.

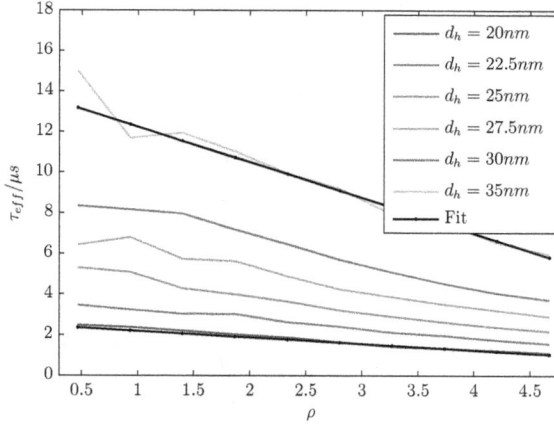

Fig. 5.2: The effective relaxation time obtained by fitting (5.14) to the magnetization response of a Brownian particle with $d_c = 20\,\text{nm}$ and different hydrodynamic diameters to a suddenly switched on magnetic field with different field strength. The linear dependency can be nicely approximated by (5.15) (Fit).

(5.9) and (5.17) can be solved with a standard ordinary differential equation solver. Subsequently, the matlab function **ode23s** is used. Only steady state solution of (5.9) and (5.17) are taken into consideration. Both equations reach a steady state region. This can be seen by deriving the general solution of

$$\frac{\mathrm{d}M}{\mathrm{d}t} = -\frac{1}{\tau}\left(\mathcal{L}(t) - M\right) \qquad (5.18)$$

by separation of variables. The solution is given by

$$M(t) = M_0 \exp(-t/\tau) + \left(\frac{1}{\tau}\int_0^t \mathcal{L}(t')\exp\left(t'/\tau\right)\mathrm{d}t'\right)\exp\left(-t/\tau\right) \quad , \qquad (5.19)$$

whereas the initial condition $M(0) = M_0$ has been used. The integral in parenthesis can then be integrated by parts

$$\frac{1}{\tau}\int_0^t \mathcal{L}(t')\exp\left(t'/\tau\right)\mathrm{d}t' = \mathcal{L}(t)\exp(t/\tau) - \mathcal{L}(0) - \int_0^t \mathcal{L}^{(1)}(t')\exp\left(t'/\tau\right)\mathrm{d}t' \quad ,$$

where $\mathcal{L}^{(1)}$ denotes the first derivative of the Langevin function. A further integration by parts leads to

$$\frac{1}{\tau}\int_0^t \mathcal{L}(t')\exp\left(t'/\tau\right)\mathrm{d}t' = \mathcal{L}(t)\exp(t/\tau) - \mathcal{L}(0) -$$
$$\tau\mathcal{L}^{(1)}(t)\exp(t/\tau) + \tau\mathcal{L}^{(1)}(0) + \tau\int_0^t \mathcal{L}^{(2)}(t')\exp(t'/\tau)\mathrm{d}t' \quad .$$

When iterating this procedure, the general solution (5.19) can be expressed in terms of

$$M(t) = M_0 \exp(-t/\tau) + \sum_{n=0}^{\infty}\left((-\tau)^n \mathcal{L}^{(n)}(t) - (-\tau)^n \mathcal{L}^{(n)}(0)\exp\left(-t/\tau\right)\right) \quad . \qquad (5.20)$$

Thus, for time intervals much greater than τ the solution is solely given by

$$\underset{t \gg \tau}{M}(t) = \sum_{n=0}^{\infty} (-\tau)^n \mathcal{L}^{(n)}(t) \quad . \tag{5.21}$$

This is the steady state solution, which does not depend on the initial conditions any more.

For example, to simulate the magnetization response of an MPS with a base frequency of $f_0 = 25\,\text{kHz}$, it is sufficient to just simulate three periods and only use the last period, which is the steady solution.

This ODE-based particle model seems to be promising, since its derivation is quite intuitive and it is possible to extend this model. For example, it would be interesting not to use the Langevin function but the equilibrium function of rigid particles with randomly orientated easy axis [28], since it is very likely that the Néel diffusion dominates the Brownian diffusion. However, this equilibrium function additionally depends on the anisotropy of the particles and therefore the magnetization response would depend on three parameters (d_c, τ, K).

Furthermore, it seems to be possible to include an anisotropic behavior by considering the relaxation time τ as a function of the scalar product $\boldsymbol{HM}$ of the applied magnetic field $\boldsymbol{H}$ and the magnetization of the particels $\boldsymbol{M}$.

5.4 Fitting Algorithm

The fitting algorithm provides a link between the experimental data and the approximated simple particle models. This enable us to extrapolate MPS experimental data to any arbitrary MPI system.

Unfortunately, it is not an easy task to find the correct distribution $\mathcal{P}(d_c, \tau_0)$ of (5.16) which models the experimental data. The distribution $\mathcal{P}(d_c, \tau_0)$ should be smooth and strict positive. Furthermore, it is indeed a two-dimensional function, since a specific particle with a specific diameter should be allowed to incorporate two or even more relaxation time constants, see(5.3). The response of the combined rotation (4.4) to a suddenly applied magnetic field consists basically of two steps. The first step is solely driven by the Néel diffusion and the second step is dominated by the Brownian diffusion. This process might be approximated with the ODE-based particle model by

$$M(t) = \mathcal{L}(\rho) \left[\frac{1}{2}(1 - e^{-t/\tau_1}) + \frac{1}{2}(1 - e^{-t/\tau_2}) \right] \quad , \tag{5.22}$$

and therefore with one specific diameter and two relaxation times.

A good way to overcome this problem is to take advantage of biological optimization strategies [54] in terms of a genetic algorithm. An ordinary genetic algorithm can only handle discrete variables, but Gutowski [26] proposed a one-dimensional smooth

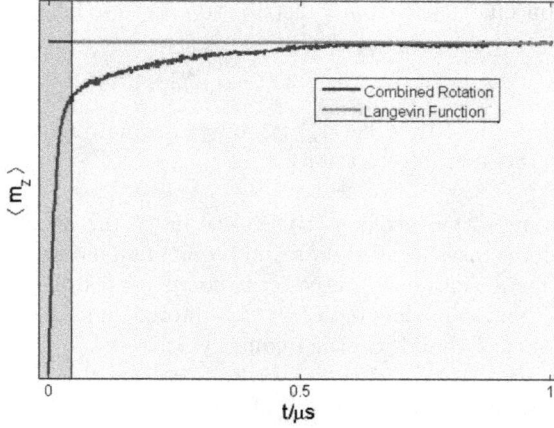

Fig. 5.3: Response magnetization of the combined rotation model. Within the blue area the process is driven by the Néel diffusion and the green region is dominated by the Brownian diffusion.

genetic algorithm. Like any other genetic algorithm, it is also based on mutation-, selection- and cross-over operations, but it is able to handle real smooth functions. The basic ideas of the one-dimensional smooth genetic algorithm have been extended in this work to a two-dimensional one and used to search the optimum distribution $\mathcal{P}(d_c, \tau_0)$.

In the beginning it is necessary to discretize the variables d_c and τ_0, whereby the indices c and 0 are omitted subsequently. Here it is important, that not an equal spaced discretization is applied. The particle diameter should be discretized according to

$$d(i) = \frac{d_{\text{start}} - d_{\text{end}}}{1 - \sqrt{N_{\text{d}}}} \sqrt{i} + d_{\text{start}} \quad \text{with} \quad i = 1 \ldots N_{\text{d}} \quad , \tag{5.23}$$

whereas d_{start} is the smallest and d_{end} the largest diameter, which are considered. N_{d} is the number of discretization points. Reasonable values which have been used are $d_{\text{start}} = 3\,\text{nm}$, $d_{\text{end}} = 35\,\text{nm}$ and $N_{\text{d}} = 30$. This non-linear discretization is based on the fact that mainly large particles contribute to the MPI signal and therefore it is good to apply a finer discretization, if the particles are big. In contrary, the discretization of τ should be finer, if τ is small. This leads to an equally spaced width of the hysteresis loops. Hence the following discretization is applied:

$$\tau(j) = \frac{N_\tau^2 \tau_{\text{start}} - \tau_{\text{end}}}{N_\tau^2 - 1}(1 - j^2) + j^2 \tau_{\text{start}} \quad \text{with} \quad j = 1 \ldots N_\tau \tag{5.24}$$

Here τ_{start} is the smallest, τ_{end} the largest value of τ and N_τ gives the number of discretization intervals. Reasonable values are $\tau_{\text{start}} = 5 \cdot 10^{-9}\,\text{s}$, $\tau_{\text{end}} = 1 \cdot 10^{-5}\,\text{s}$ and $N_\tau = 40$.

Furthermore, the continuous probability distribution needs to be switched to a discrete one

$$\mathcal{P}(d_c, \tau_0) \rightarrow \mathcal{W}(i,j) \quad \text{with} \quad i = 1 \ldots N_d \quad ; j = 1 \ldots N_\tau \quad , \tag{5.25}$$

where it has to be verified, that $\mathcal{W}(i,j)$ is always normalized and therefore fulfills $\sum_{(i,j)} \mathcal{W}(i,j) = 1$ throughout all iterations.

Every optimization algorithm needs a criterion to judge the obtained solution. The comparison between the measurement data and the simulated data should be evaluated in the frequency domain, since here it is easier to compensate differences of the particle concentration (5.27) and much less data has to be stored, since it is only necessary to consider the harmonics of the MPS base frequency.

The algorithm starts by calculating the steady state solutions of (5.17) for all different diameters $d(i)$ and relaxation times $\tau(j)$. Hereafter the time series get Fourier-transformed are and stored in a complex, three-dimensional matrix $S_n^{sim}(i,j)$, so that these calculations do not need to be repeated. i and j denote the spectrum of the ith diameter and the jth relaxation time and n stands for the nth harmonic. The range of n is $n = 1 \ldots N_h$, whereas N_h is the number of all considered harmonics. Since the MPS of the Institute of Medical Engineering in Luebeck stores a corrected measured spectrum S_n^{meas}, it is not necessary to perform an additional time derivative. In general, this needs to be done, since the induced signal is influenced by the time derivative of the magnetization (2.5). It is also not necessary to include effects of the sensitivity of the induction coil, since this has been adjusted by calibration.

The spectrum then needs to be calculated, which is predicted by $\mathcal{W}(i,j)$. This is done by the following multiplication

$$\sum_{(i,j)} \mathcal{W}_k(i,j) S_n^{sim}(i,j) = S_{n,k}^{sim} \quad . \tag{5.26}$$

The additional index k has been introduced, since subsequently more than one probability distribution have to be taken into consideration. Up to now the weighted, simulated spectrum $S_{n,k}^{sim}$ is so to say the spectrum of a sample with a particle concentration of $c_N = 1/V_s$. In order to determine the error function, the number of particles that are present in the ferrofluid have to be estimated. This can be done by

$$c_k = \frac{1}{N_h} \sum_{n=1}^{N_h} \log_{10} |S_n^{meas}| - \log_{10} |S_{n,k}^{sim}| \quad . \tag{5.27}$$

The estimated number of particles of the kth distribution is then given by $N_k = 10^{c_k}$ (5.16). In general, it is important to apply the logarithm first, before comparing the simulated against the measured spectrum. This is caused by the exponential decay of the power of the higher harmonics.

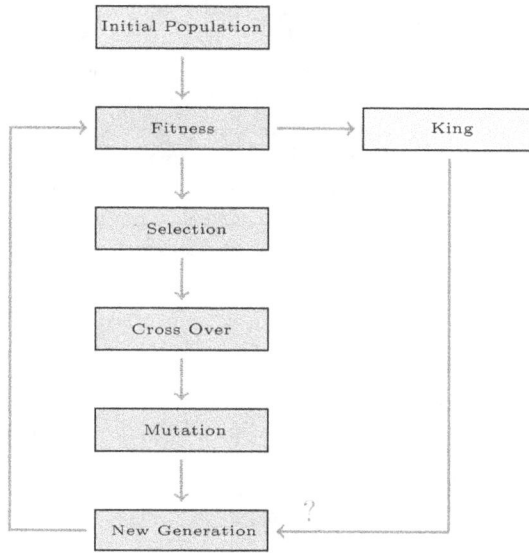

Fig. 5.4: Flowchart of the genetic algorithm. The red cycle indicates the cycle, which is repeated from generation to generation.

The error function is a composition of the differences between the power spectra and the imaginary part of the spectra

$$\mathcal{E}_k = \sum_{n=1}^{N_h} \left[\left(\log_{10}(|S_n^{\text{meas}}|) - (\log_{10}(|S_{n,k}^{\text{sim}}|) + c_k) \right)^2 \right.$$
$$\left. + \left(\log_{10}(|\text{Im}(S_n^{\text{meas}})|) - (\log_{10}(|\text{Im}(S_{n,k}^{\text{sim}})|) + c_k) \right)^2 \right] \quad , \tag{5.28}$$

where once again a logarithmic scaling is applied.

The actual smooth genetic algorithm begins by defining an initial population, which evolves from generation to generation. Each generation-cycle is separately affected by selection, cross-over and mutation operations. The initial start distributions are smooth two-dimensional Gaussian functions

$$\mathcal{W}_k(i,j) = A \exp\left(-\frac{1}{2} \left((i - \mu_{\text{d}}(k))^2 / \sigma_{\text{d}}^2 + (j - \mu_\tau(k))^2 / \sigma_\tau^2 \right) \right)$$
$$k = 1 \ldots N_{\text{ind}} \quad . \tag{5.29}$$

Here the variance is taken to be $\sigma_{\text{d}}^2 = \sigma_\tau = 2$. Furthermore, the mean values μ_{d} and μ_τ have to be selected in such a way that the N_{ind} individual probability distributions $\mathcal{W}_k(i,j)$ are equally spread over the surface defined by $N_{\text{d}} \times N_\tau$. $N_{\text{ind}} \approx 100$ has been found to be a sufficiently large population.

After the initial population has been created, the fitting algorithm enters a cycle, which runs over the generations, see figure (5.4). It starts by evaluating the fitness of the N_{ind} individual probability distributions. Here the error function (5.28) is used as an indicator of fitness. The specific individual probability distribution with the smallest error is automatically saved, so that it cannot be destroyed by the following mutation or cross-over operations. This specific probability distribution is called the King (see (5.4), right red loop). If the King distribution function is destroyed, it will re-enter the population and replaces the weakest individual. This is of course a non-biological behavior of the algorithm, but it ensures that the error of the best individual is a monotonically decreasing function.

The evaluation step is then followed by the selection step. Within the selection step the decision is made, whether an individual is allowed to become a parent and thereby pass its characteristics on to the next generation. According to Gutowski, the decision rule

$$r < 1/\left(1 + \exp(\mathcal{E}_k - \frac{M}{s})\right) \tag{5.30}$$

is used, where M is the mean error and s is the standard deviation, defined by

$$M = \frac{1}{N_{\text{ind}}} \sum_{k=1}^{N_{\text{ind}}} \mathcal{E}_k$$
$$s = \sqrt{\frac{1}{N_{\text{ind}}} \sum_{k=1}^{N_{\text{ind}}} (\mathcal{E}_k - M)^2} \quad . \tag{5.31}$$

r is an equal distributed random number of the interval $[0,1]$. If this inequality holds, the kth individual can become a parent and therefore take part in the cross-over operation.

The reproduction or cross-over operation is performed by the following equations, see figure (5.5).

$$O_1 = P_1(i,j) \cdot (1 - \text{Co}_l(i,j)) + P_2(i,j)\text{Co}_l(i,j)$$
$$O_2 = P_2(i,j) \cdot \text{Co}_l(i,j) + P_2(i,j) \cdot (1 - \text{Co}_l(i,j)) \tag{5.32}$$

P_1 and P_2 are the parents-distributions. They are two different distribution functions $\mathcal{W}_k(i,j)$, which both have passed the selection process (5.30). O_1 and O_2 are the new offspring-distributions, which replace the old parents-distributions. The mixing

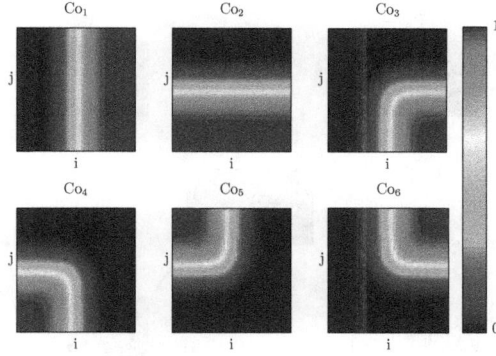

Fig. 5.5: Cross-over functions, calculated with $N_d = N_\tau = 50$ and fixed mean values $\mu_d^{co} = \mu_\tau^{co} = 25$.

or cross-over functions $Co(i, j)$ are smooth functions, which are given by:

$$Co_1(i, j) = \frac{1}{2}(1 + \tanh((i - \mu_d^{co})/2)) \quad \forall j$$

$$Co_2(i, j) = \frac{1}{2}(1 + \tanh((j - \mu_\tau^{co})/2)) \quad \forall i$$

$$Co_3(i, j) = \frac{1}{2}(1 + \tanh((i - \mu_d^{co})/2) \cdot \frac{1}{2}(1 + \tanh((j - \mu_\tau^{co})/2))$$

$$Co_4(i, j) = \frac{1}{2}(1 - \tanh((i - \mu_d^{co})/2) \cdot \frac{1}{2}(1 + \tanh((j - \mu_\tau^{co})/2))$$

$$Co_5(i, j) = \frac{1}{2}(1 - \tanh((i - \mu_d^{co})/2) \cdot \frac{1}{2}(1 - \tanh((j - \mu_\tau^{co})/2))$$

$$Co_6(i, j) = \frac{1}{2}(1 + \tanh((i - \mu_d^{co})/2) \cdot \frac{1}{2}(1 - \tanh((j - \mu_\tau^{co})/2))$$

These weighting functions are always in the interval $[0, 1]$, see figure (5.5). It is randomly determined, which cross-over function is applied, where each function has the same probability to be chosen. The value of μ_d is a random value out of the interval $[1+2, N_d-2]$ and likewise μ_τ is randomly chosen value out of the interval $[1+2, N_\tau-2]$. This lengthy procedure of creating the cross-over function ensures that each characteristic of a parent-distribution is able to contribute to the new offspring-distribution.

The cross-over operation mixes only old properties of the population, therefore it is important to additionally introduce new features to the population. This is done by mutation operations, which are divided into local and global mutations. Each individual probability distribution has the same probability to be mutated. This probability for mutation has been chosen in that way that in average ten individuals per generation cycle undergo a mutation operation. The local mutation is done by the following operation

$$\mathcal{W}_k(i, j) = L^{\pm}(i, j) \cdot \mathcal{W}_k(i, j) \quad , \tag{5.33}$$

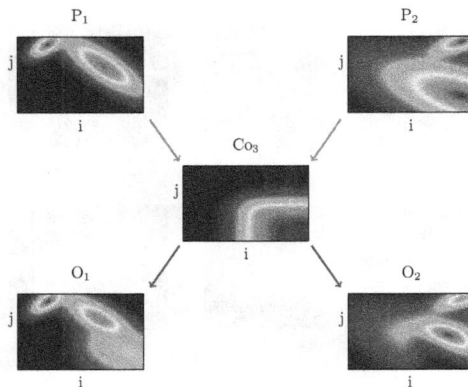

Fig. 5.6: The cross-over operation between two arbitrary parent-distributions produces two new offspring-distributions.

whereas $L^\pm(i,j)$ is given by the general two-dimensional Gaussian distribution

$$L^\pm(i,j) = 1 \pm \frac{1}{2} \exp\left(-\frac{1}{2(1-\kappa^2)}\left[\frac{(i-\mu_d)^2}{\sigma_d^2} + \frac{(j-\mu_\tau)^2}{\sigma_\tau^2} - \frac{2\kappa(i-\mu_d)(j-\mu_\tau)}{\sigma_d\sigma_\tau}\right]\right) .$$
$$(5.34)$$

Whether the $+$ or $-$ function is applied is again determined randomly. The $+$ function enhances the values of $\mathcal{W}_k(i,j)$ around the point (μ_d, μ_τ) and the $-$ function decreases the values, see figure (5.7). The parameters σ_d, σ_τ, μ_τ, μ_d and the correlation coefficient κ are also not constant, but randomly chosen from the intervals

$$\sigma_d, \sigma_\tau \in [1,4]$$
$$\mu_d \in [2, N_d - 1]$$
$$\mu_\tau \in [2, N_\tau - 1]$$
$$\kappa \in [-0.7, 0.7] .$$

The global mutation is performed by the following function

$$\mathcal{W}_k(i,j) = G^\pm(\mathcal{W}_k(i,j)) = \exp\left(\pm\left[\mathcal{W}_k(i,j)/\max_{(i,j)}(\mathcal{W}_k(i,j)) - \frac{1}{2}\right]\right) . \qquad (5.35)$$

Likewise to the local mutation a random number is used to determine, whether the $+$ or the $-$ function should be applied. The $+$ function increases every value of $\mathcal{W}_k(i,j)$, which fulfills the condition $\mathcal{W}_k(i,j)/\max_{(i,j)}(\mathcal{W}_k(i,j)) > \frac{1}{2}$, and reduces the remain. Contrary to this the $-$ function flattens the probability distribution, see figure (5.8).

Again, a random selection process determines, whether only the local or the global mutation or even both reshape the considered individual distribution.

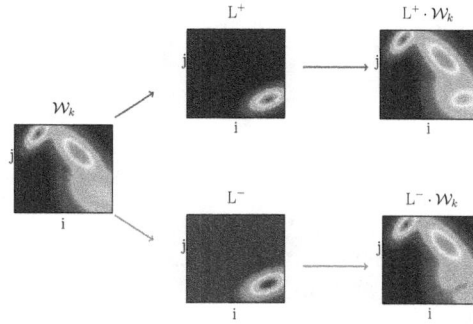

Fig. 5.7: Local mutation operation applied to an arbitrary function $\mathcal{W}_k$. L^+ inserts a maximum and L^- inserts a minimum.

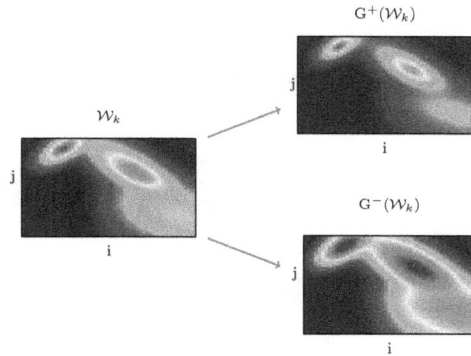

Fig. 5.8: Global mutation operation applied to an arbitrary function. G^+ enhances the difference between maxima and minima and G^- flattens the distributions.

The generation cycle finally stops, if there has been no improvement of the King solution over the last 200 generations or if the number of generation reaches the limit of 1500. This two-dimensional genetic algorithm may be of course applied to any different optimization problem. Since it incorporates many parameters, which for example control the number of mutations per generation cycle or the size of the population, there is a big potential to use this algorithm, even for very special optimization problems.

6 Simulation of a Frequency Sweep

The stochastic particle model may now be used to investigate lots of different questions such as:

- What is the best trajectory?

- How should a particle look like to detect molecular binding processes?

- How should a particle be designed to perform hyperthermia?

- Would it be interesting to build a MPS that is able to apply different field directions?

- ...

Another interesting questions is "Is there an optimal base frequency?".

The base frequency is often chosen to be 25 kHz. This choice may be justified by two practical reasons. On the one hand this frequency is out of the human audible range and on the other hand this frequency can still be amplified by typical audio-amplifiers. Assuming that the particle dynamic is not affected by the choice of the applied frequency, the induced voltage (2.3) can be calculated by

$$u(t) \quad = \quad -\mu_0 \int_\Omega \partial_t \boldsymbol{M}(\boldsymbol{H}_{\text{ext}}(\boldsymbol{r}, t), \boldsymbol{r}, t) \boldsymbol{p}(\boldsymbol{r}) \mathrm{d}r^3 \tag{6.1}$$

$$\stackrel{\text{static}}{=} \quad -\mu_0 \omega \int_\Omega \boldsymbol{H}_0 \cos(\omega t) \frac{\mathrm{d}\boldsymbol{M}(\boldsymbol{H}_{\text{ext}})}{\mathrm{d}\boldsymbol{H}_{\text{ext}}} \boldsymbol{p}(\boldsymbol{r}) \mathrm{d}r^3 \quad , \tag{6.2}$$

whereas the external field is given by $\boldsymbol{H}_{\text{ext}} = \boldsymbol{H}_0 \sin(\omega t)$. Thus, this estimation predicts a linear rise of the SNR with respect to the base frequency. It is therefore of interest, whether this estimation is correct or whether the particle dynamic behaves differently. Furthermore, it is important to note that time-varying magnetic fields induce eddy currents in a biological tissue. These currents may cause harm to the human body, since they lead to local temperature increase or to nerve stimulations. This effect is measured by the specific absorption rate (SAR), which is the amount of energy deposition per weight of the tissue [8, 23]. The SAR is a quadratic function of the applied frequency of the magnetic field and therefore it is impossible to use arbitrary high frequencies for MPI. A reasonable base frequency range is from 20 kHz to 100 kHz [23].

To obtain a good overview of the particle dynamic, it is necessary to simulate different particles with different core diameters d_{c}, different hydrodynamic diameters d_{h} and different anisotropy values K. Furthermore, it has been examined that it is indispensable

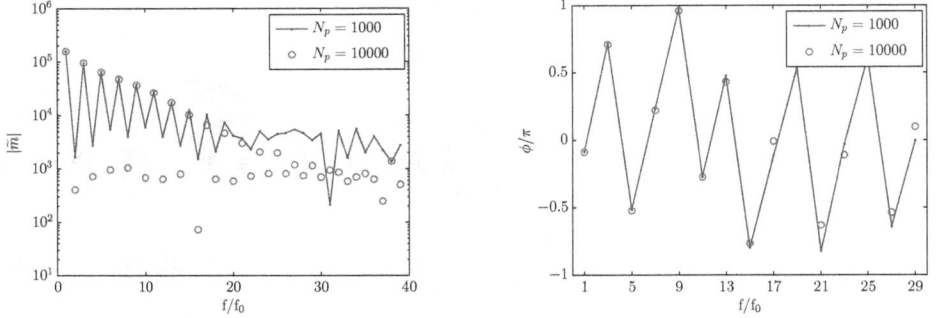

Fig. 6.1: A comparison between two particle signals, which have been simulated with the same parameters, but with different number of particles N_{p}. Used simulation paramters: $T = 300\,\mathrm{K}$, $H = 0.015\,\mathrm{mT}/\mu_0$, $d_{\mathrm{c}} = 20\,\mathrm{nm}$, $d_{\mathrm{h}} = 30\,\mathrm{nm}$, $K = 1 \cdot 10^4 \mathrm{J/m^3}$.

to simulate at least three periods of each frequency to obtain the steady state solution. Therefore, this simulation study is computationally expensive and it is interesting to know how many particles N_{p} have to be simulated. This has been tested by simulating time series with $N_{\mathrm{p}} = 10000$ and with $N_{\mathrm{p}} = 1000$ particles and comparing both results, see figure (6.1). It has been found that the solutions strongly depend on the ratio of the Zeeman energy to the thermal energy (figure (3.8)) and on the anisotropy K of the particles. For example often only the third harmonic of $d_{\mathrm{c}} = 15\,\mathrm{nm}$-particles simulated with $N_{\mathrm{p}} = 1000$ is trustable, whereas it is indeed possible to gain information out of the 7th harmonic of $d_{\mathrm{c}} = 20\,\mathrm{nm}$-particles, which have been simulated with $N_{\mathrm{p}} = 1000$. However, it is sufficient to simulate only $N_{\mathrm{p}} = 1000$.

Throughout the simulation, the reduced magnetic moment $\boldsymbol{m} = |\boldsymbol{\mu}|/\mu$ has been considered, since only the particle dynamic is of interest. In addition, the magnetic field has been assumed to be homogenous and fixed in the z direction and thus only the z component $m = \boldsymbol{m} \cdot \boldsymbol{e}_z$ is important. The simulated data consists of noisy, discrete time series and therefore the time derivative should not be performed in the time domain but in the frequency domain, according to

$$\widetilde{m} = \mathcal{F}\left(\frac{\mathrm{d}m}{\mathrm{d}t}\right) = i\omega\mathcal{F}\left(m\right) \quad . \tag{6.3}$$

Additional parameters, which have been used are presented in table (6.1).

In the figures (6.2), (6.3) and (6.4) some results of the frequency sweep are presented. They show the amplitude and the phase of the third harmonic. For reference, each figure also presents the solution of a superparamagnetic particle. This solution depends solely on the considered particle diameter and is simulated by the Langevin equation (3.46).

Furthermore, the figure (6.5) presents the ratio between the amplitudes of the fifth and the third harmonic $|\widetilde{m}_5|/|\widetilde{m}_3|$, which is an indicator for the slope of the spectrum.

Table 6.1: Simulation Parameters

Name	Symbol	Value
Temperature	T	$300\,\mathrm{K}$
Magnetic Field Strength	H	$0.015\,\mathrm{mT}/\mu_0$
Saturation Magnetization Magnetite	M_{sat}	$0.6\,\mathrm{mT}/\mu_0$
Viscosity of Water	$\eta_{\mathrm{H_2O}}$	$0.9 \cdot 10^{-3}\,\mathrm{Ns/m^2}$
Néel Damping Constant	α	0.1

In addition, some frequency sweeps have been simulated by taking only the Néel or the Brownian diffusion into account, whereas the Néel diffusion process has been simulated with particles, whose easy axis are randomly orientated. Some results are presented in figure (6.6) and figure (6.7).

For analyzing its frequency behavior, a frequency sweep of the ODE-based particle model has also been simluated. This result is presented in figure (6.8).

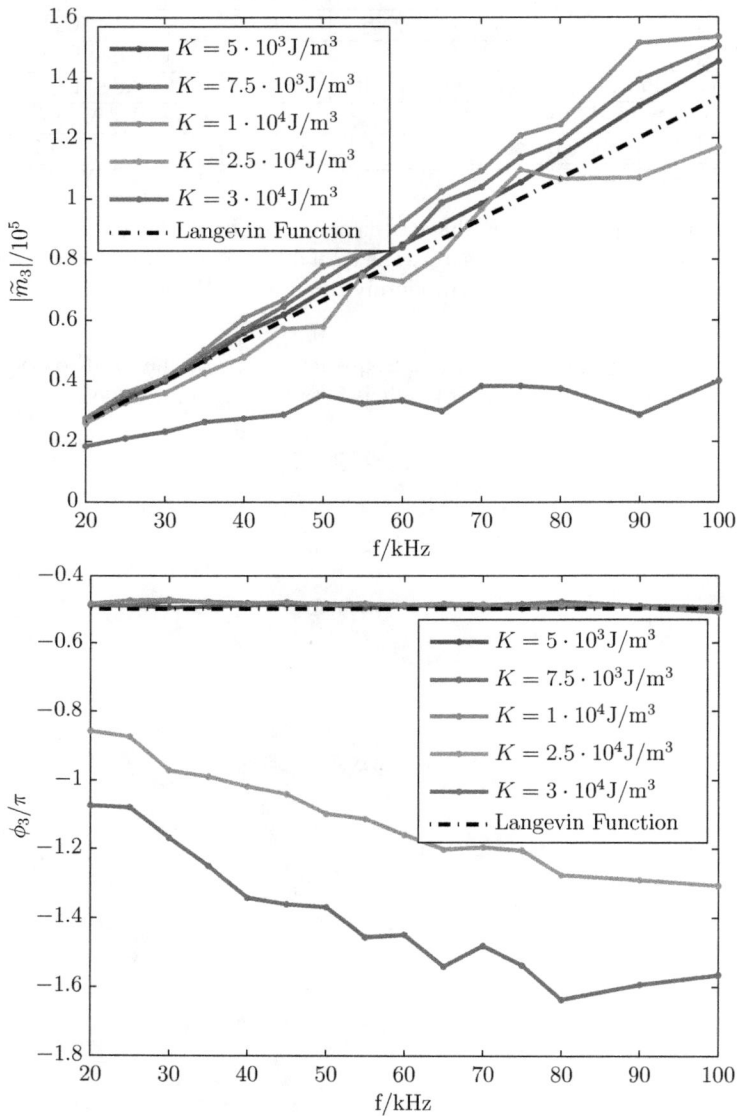

Fig. 6.2: Amplitude and phase of the third harmonic against frequency of a particle specified by $d_c = 15\,\mathrm{nm}$, $d_h = 25\,\mathrm{nm}$ and different anisotropy values.

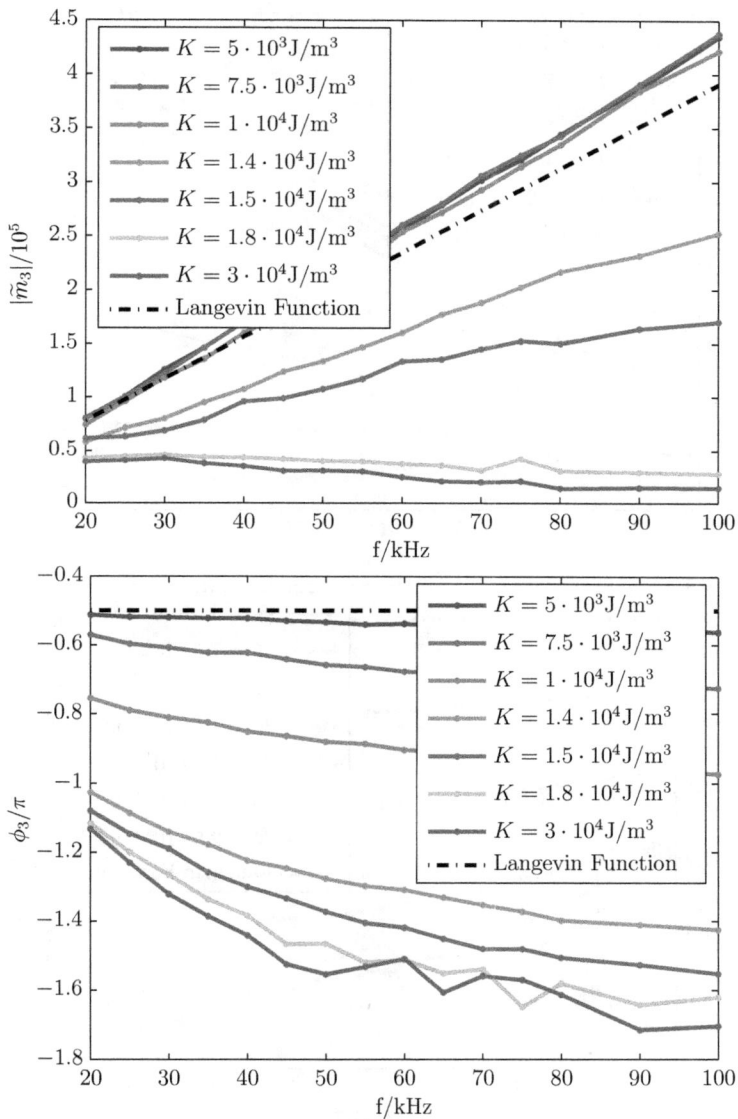

Fig. 6.3: Amplitude and phase of the third harmonic against frequency of a particle specified by $d_c = 20\,\text{nm}$, $d_h = 30\,\text{nm}$ and different anisotropy values.

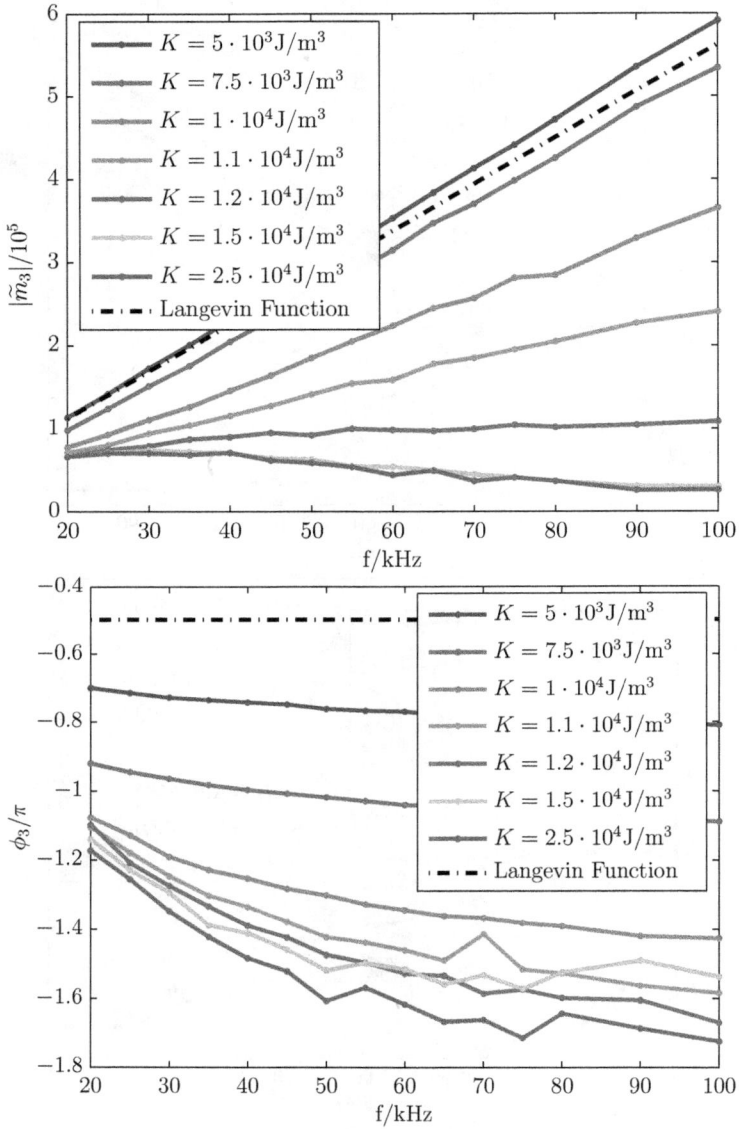

Fig. 6.4: Amplitude and phase of the third harmonic against frequency of a particle specified by $d_c = 25\,\mathrm{nm}$, $d_h = 35\,\mathrm{nm}$ and different anisotropy values.

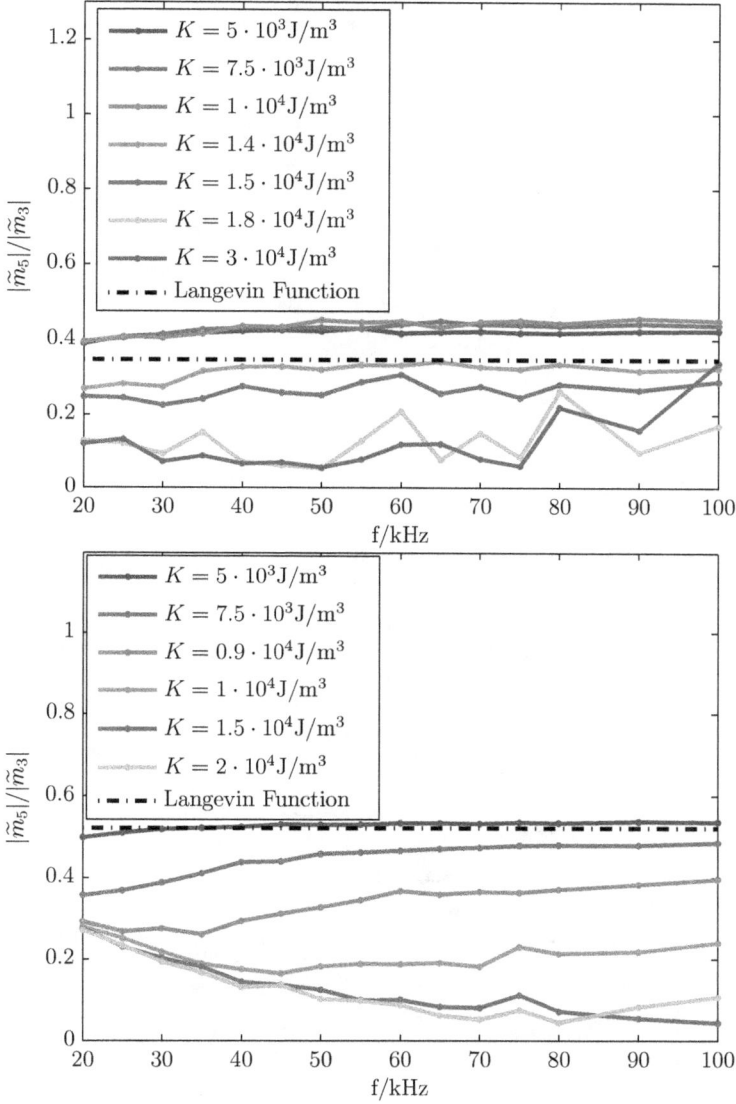

Fig. 6.5: Ratio between the amplitudes of the fifth and the third harmonic $|\widetilde{m}_5|/|\widetilde{m}_3|$. The figure at the top is simulated with $d_\mathrm{c} = 20\,\mathrm{nm}$, $d_\mathrm{h} = 30\,\mathrm{nm}$ particles and the figure at the bottom is simulated with particles specified by $d_\mathrm{c} = 30\,\mathrm{nm}$, $d_\mathrm{h} = 40\,\mathrm{nm}$.

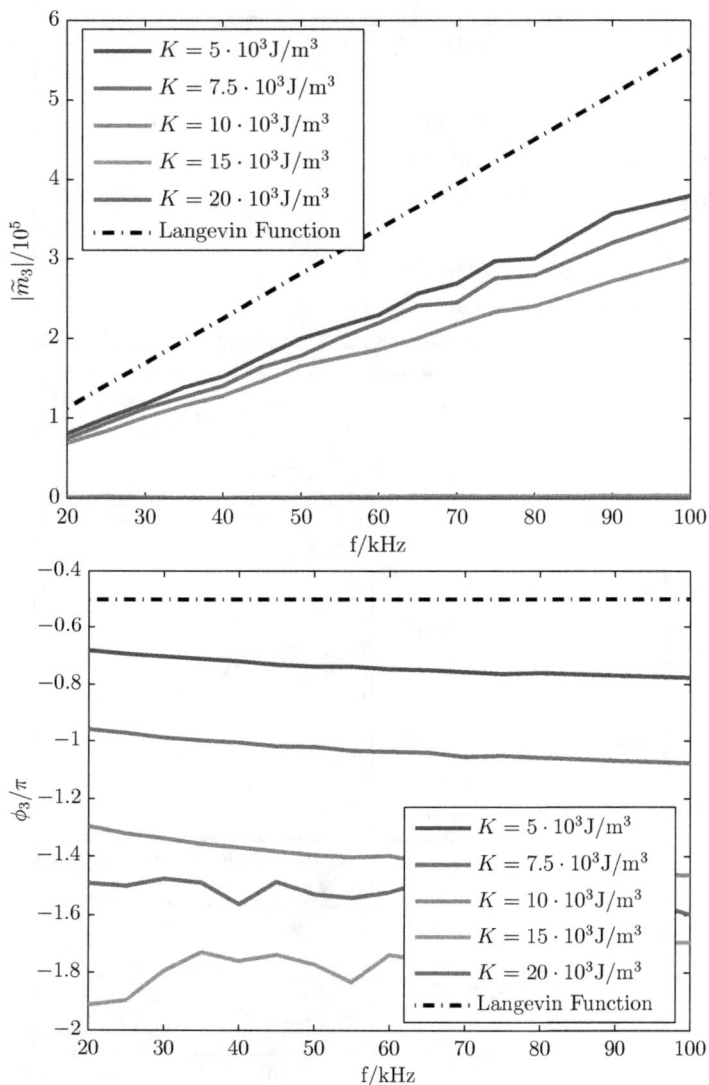

Fig. 6.6: Amplitude and phase of the third harmonic against frequency of a particle with randomly orientated easy axis specified by $d_c = 25\,\text{nm}$. This figures show the simulation result of the Néel diffusion process of particles with randomly oriented easy axis.

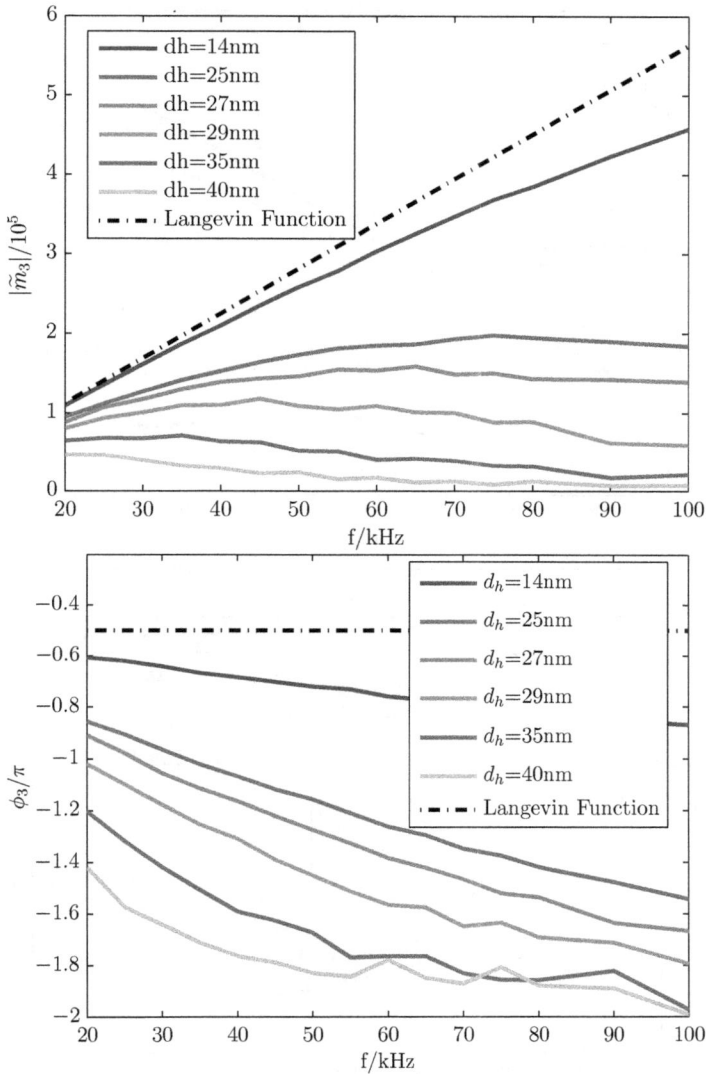

Fig. 6.7: Amplitude and phase of the third harmonic against frequency of a particle with a particle diameter of $d_c = 25\,\mathrm{nm}$. This figure has been simulated by taking only the Brownian diffusion into account.

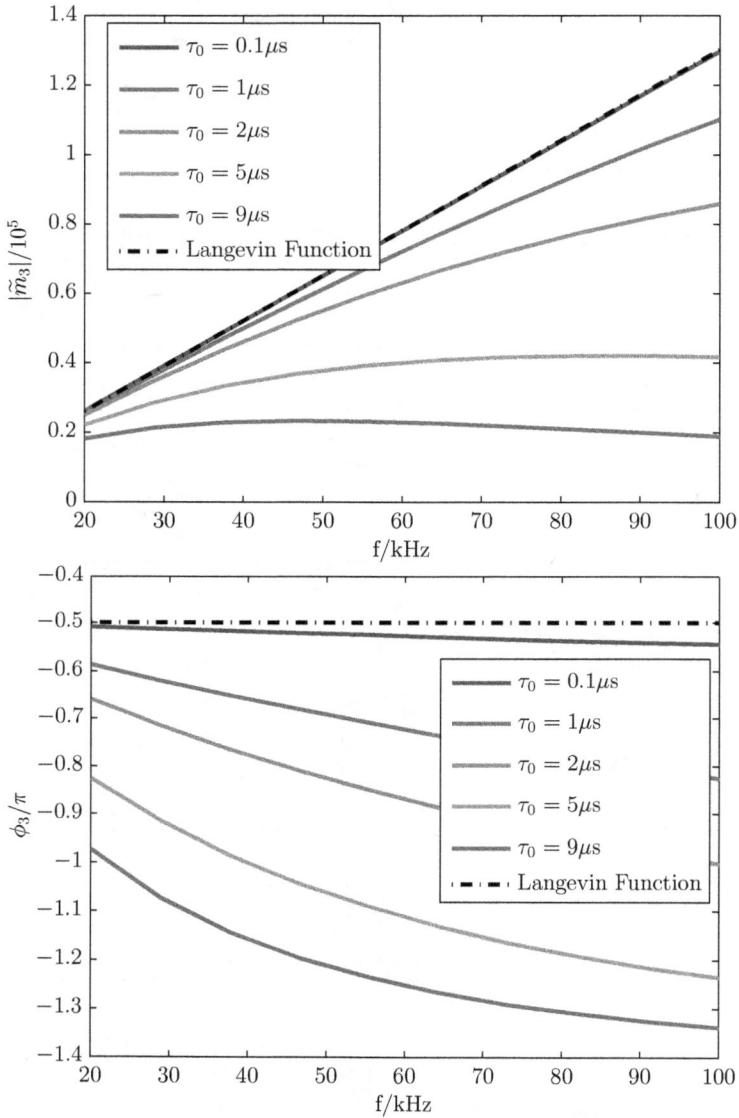

Fig. 6.8: Amplitude and phase of the third harmonic against frequency simulated with the ODE-based particle model of a particle with a diameter of $d_c = 20\,\text{nm}$.

The relation between the amplitude of the third harmonic and the frequency, which is presented in the figures (6.2), (6.3) and (6.4), is approximately linear, whereas the slope depends strongly on the anisotropy constant K. For particles with a low anisotropy it is positive and even sometimes larger than predicted by the Langevin equation. In contrast, it may also be negative, if the anisotropy constant is large.

The magnetization behavior of particles with a high anisotropy constant can be understood in terms of a blocked particle. Here, the particles are only influenced by the Brownian diffusion and their magnetization behavior coincides with that one obtained by simulating only the Brownian diffusion process, see figure (6.7). These particles are not able to contribute to the MPI signal.

Only if the Néel diffusion process is not blocked, the correlation between the amplitude of the third harmonic and the frequency is positive. However, by comparing figure (6.6) with figure (6.4), it is also apparent that the signal of particles, which are simulated by the combined rotation model, differs from the signal that is simulated by taking only the Néel diffusion into account. An essential difference is that the amplitude of the third harmonic of particles with a low anisotropy, simulated by the combined rotation model, is even larger than the amplitude of the corresponding superparamagnetic particles. This can be visualized by considering the ratio $|\widetilde{m}_3|/|\widetilde{m}_3^L|$ between the amplitudes of the third harmonic of particles, simulated by the combined diffusion process, $|\widetilde{m}_3|$ and the corresponding superparamagnetic particles $|\widetilde{m}_3^L|$. The figure (6.9) shows the ratio $|\widetilde{m}_3|/|\widetilde{m}_3^L|$ simulated with three different core diameters and a base frequency of $f = 50\text{kHz}$. It is conspicuous that the ratio $|\widetilde{m}_3|/|\widetilde{m}_3^L|$ depends strongly on the diameter and therefore on the volume of the particles. This can be understood in terms of the Stoner Wohlfarth particle model (3.3.2), since here the potential barrier is given by

$$\Delta E = K \cdot V_{\mathrm{c}} = K \cdot \frac{\pi}{6} d_{\mathrm{c}}^3 \quad . \tag{6.4}$$

Therefore, the product of the anisotropy constant K and the particle volume determines, whether a particle is blocked or not. Furthermore, by considering the curve of the $d_{\mathrm{c}} = 15\text{nm}$ particle, it is interesting to note that if the anisotropy constant decreases the ratio $|\widetilde{m}_3|/|\widetilde{m}_3^L|$ first increases and reaches a maximum, which is greater than 1. If the anisotropy is further reduced, the ratio $|\widetilde{m}_3|/|\widetilde{m}_3^L|$ starts to decrease. It is to be expected that $|\widetilde{m}_3|/|\widetilde{m}_3^L|$ converges to 1, since the particle can be assumed to be superparamagnetic. This transition to the superparamagnetic state can be also identified by considering the phase lack of the $d_{\mathrm{c}} = 15\text{nm}$ particle, see figure (6.2). If the anisotropy constant is smaller than $K \approx 1 \cdot 10^4 \, \text{J/m}^3$, the magnetization response is in-phase with a superparamagnetic particle. Moreover, the big influence of the anisotropy on the particle signal can be studied by considering the ratio $|\widetilde{m}_5|/|\widetilde{m}_3|$ between the amplitudes of the fifth and the third harmonic, which is presented in figure (6.5). This ratio can serve as an indicator for the decrease of the spectrum. Since MPI needs to detect as many harmonics as possible, this ratio should be large. Likewise to the already presented results, the ratio $|\widetilde{m}_5|/|\widetilde{m}_3|$, and therefore also the MPI-signal, is larger, if the anisotropy of the particles is small.

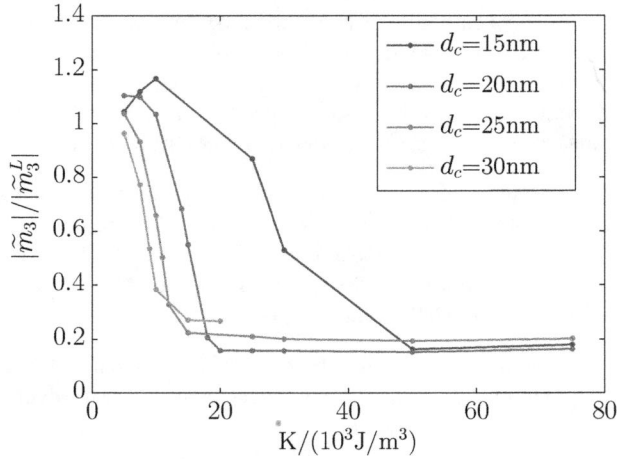

Fig. 6.9: Ratio $|\widetilde{m}_3|/|\widetilde{m}_3^L|$ between the amplitudes of the third harmonic of particles with different core diameters that are simulated with the combined diffusion process and the corresponding superparamagntic particles. Each particle is coated by a 5 nm thick shell and the ratio $|\widetilde{m}_3|/|\widetilde{m}_3^L|$ is evaluated at the frequency of $f = 50\,\text{kHz}$.

The result of the frequency sweep, which has been simulated by the ODE-based particle model (see figure (6.8)), is promising. The ODE-based particle model exhibits equal properties compared to the Brownian or the Néel diffusion (figure (6.7) and figure (6.6)). As expected, it coincides with the Langevin function in the limit of a small relaxation time τ_0 and in the limit of a large relaxation time τ_0 the phase lack converges to the maximum of $\Delta\Phi = -\pi$.

It is evident from this simulation study that the anisotropy of the particles has a significant influence on the MPI signal. Furthermore, the anisotropy values of the particles, which are able to contribute to the MPI signal, are quite small. They are in the lower range or even below the range, which is in general assumed for magnetite particles in literature ($K \approx 1 \cdot 10^4 \cdots 1 \cdot 10^5\,\text{J/m}^3$) (3.2.3). This can be understood in terms of non-coherent magnetic reversal modes (3.3.2). In addition, this simulation study shows that the magnetization behavior of the contrast agent is mainly driven by the Néel diffusion and only modulated by the Brownian diffusion. This coincides with the results obtained by J. Weizenecker et al [33].

Due to the found linear correlation between the MPI signal and the frequency in the range from 20kHz to 100kHz, with a positive or a negative slope, it is not clear, how the particle signal depends on the frequency. However, it is very likely that a contrast agent, which shows a good performance at a base frequency of 25kHz will also generate a good MPI signal at a higher frequency.

7 Demagnetization Effects and Particle Interactions

The basic assumption (2.11) that the magnetization of a ferrofluid increases linearly with the particle concentration c_N. This is valid, if effects which depend on particle concentration or on sample volume, are negligible.

To measure the particle concentration c_N, it is common use to switch to the iron concentration c_{Fe}. In the case of magnetite nanoparticles the iron concentration is linked to the particle concentration due to the following equation

$$c_N = \frac{1}{3} \frac{M_{Fe_3O_4}}{\rho_{Fe_3O_4} \langle V_c \rangle} \cdot c_{Fe} \quad . \tag{7.1}$$

$M_{Fe_3O_4}$ is the molar mass of magnetite, which is given by

$$M_{Fe_3O_4} = 3M_{Fe} + 4M_O = 3 \cdot 55.85 \frac{g}{mol} + 4 \cdot 15.99 \frac{g}{mol} = 231.54 \frac{g}{mol} \quad . \tag{7.2}$$

$\rho_{Fe_3O_4} \approx 5200\,\mathrm{kg/m^3}$ stands for the mass density of magnetite and $\langle V_c \rangle$ is the averaged core volume of the particles. c_{Fe} overestimates c_N, since not all particles contribute to the signal. One may think of a little fraction of particles, which are either too large or too small to have a constant magnetic moment (3.2.4) or of particles, which have a large anisotropy (6.1). Furthermore, the iron concentration is an important parameter with respect to medical applications, since a high iron dose can cause toxic side effects [49].

Thinking of medical applications the iron concentration is in general very low. For example, a bolus injection of $1\,\mathrm{ml}$ of a contrast agent with $c_{Fe} = 0.5\,\mathrm{mol/l}$ leads to an iron concentration within the blood, which is below $100\,\mu\mathrm{mol/l}$, supposed that the whole contrast agent remains in the human blood pool ($6 - 71$). Here particle interactions and demagnetization effects can be safely neglected. But as described in (2.3) the concentration of the contrast agent should be as high as possible, while measuring an MPI system function. Therefore, a rough estimation is made, up to which concentration the linear relationship between the magnetization and the particle concentration is justifiable.

Demagnetization Effects

Analogue to the demagnetization field, which induces the shape anisotropy of nanoparticles (3.2.3), a demagnetization field has to be taken into consideration, if a ferrofluid

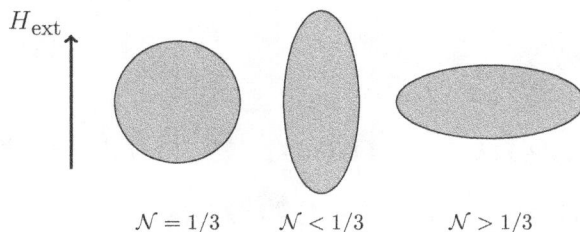

$$\mathcal{N} = 1/3 \qquad \mathcal{N} < 1/3 \qquad \mathcal{N} > 1/3$$

Fig. 7.1: The demagnetization factor depends on the shape of the sample and its direction to the external magnetic field. The demagnetization effect vanishes if the sample is very narrow and points in the direction of the external field.

sample is placed in a magnetic field [3, 43]. This demagnetization field always points in opposite direction to the external magnetic field and therefore reduces the magnetic field in the sample. The magnetic field $\boldsymbol{H}_{\text{in}}$ in the sample can be calculated by

$$\boldsymbol{H}_{\text{in}} = \boldsymbol{H}_{\text{ext}} - \mathcal{N}\boldsymbol{M} \qquad (7.3)$$

Here $\mathcal{N}$ is the demagnetization factor, which in the case of a rotational symmetric ellipsoidal sample shape can be calculated with (3.20). The magnetization of a ferrofluid sample is only effected by $\boldsymbol{H}_{\text{in}}$ and thus the static magnetization has to be calculated by the relation

$$M_{\text{demag}}(H_{\text{ext}}, T) = c_{\text{N}}|\boldsymbol{\mu}|\mathcal{L}\left(\frac{\mu_0|\boldsymbol{\mu}|(H_{\text{ext}} - \mathcal{N}M_{\text{demag}})}{k_{\text{B}}T}\right) \qquad . \qquad (7.4)$$

It is important to note that if a two- or three-dimensional system function should be measured, one should only use spherical samples. All different shapes would induce demagnetization effects, which are no longer isotropic. In contrast, the sample used in a one-dimensional MPS should be as elongated as possible, whereas the symmetry axis should be parallel to the external field. This would reduce the demagnetization field to a minimum, see Fig. (7.1).

Particle Interactions

If the particle concentration increases, particle-particle interactions get more and more important. The dipole-dipole interactions support the alignment of the magnetic moments and therefore increase the steepness of the magnetization curve.

The effect of dipole-dipole interactions can be estimated with a simple mean field model. Weiss assumed that the effective field, which is acting on a single particle, is actually the field in an empty sphere surrounded by a ferrofluid (7.2). Contrary to the demagnetization field, the strength of the magnetic field in the sphere will be enhanced due to the contribution of the surrounding particles and can be calculated by

$$\boldsymbol{H}_{\text{Weiss}} = \boldsymbol{H}_{\text{ext}} + \frac{1}{3}\boldsymbol{M} \qquad . \qquad (7.5)$$

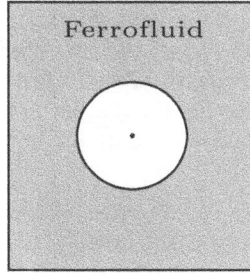

Fig. 7.2: Weiss calculated the effective field, which includes particle-particle interactions, by calculating the magnetic field, which is in an empty sphere surrounded by a ferrofluid.

Analogous to (7.4), the magnetization, which includes dipole-dipole interactions, can be calculated by

$$M_{\mathrm{dd}}(H_{\mathrm{ext}}, T) = c_{\mathrm{N}}|\boldsymbol{\mu}|\mathcal{L}\left(\frac{\mu_0|\boldsymbol{\mu}|(H_{\mathrm{ext}} + \frac{1}{3}M_{\mathrm{dd}})}{k_{\mathrm{B}}T}\right) \quad . \tag{7.6}$$

Ivanov et al investigated the influence of dipole-dipole interaction on the magnetization curve in dipolar fluids by comparing experimental data and measurement data [29]. They found that the model is correct in the case of low concentrated ferrofluids below a sample saturation magnetization of $M_{\mathrm{sat}} < 5 \cdot 10^3 \mathrm{A/m}$. Since it is very likely that the saturation magnetization of the MPI contrast agents is lower than this value (8.5), the Weiss model seems to be applicable.

The demagnetization effects and the dipole-dipole interactions have to be combined to understand the measurement of a ferrofluid sample. This can be done by introducing an effective magnetic field, which considers both effects. This effective field $\boldsymbol{H}_{\mathrm{eff}}$ is given by

$$\boldsymbol{H}_{\mathrm{eff}} = \boldsymbol{H}_{\mathrm{ext}} - (\mathcal{N} - \frac{1}{3})\boldsymbol{M} \tag{7.7}$$

and therefore the magnetization can be calculated by

$$M(H_{\mathrm{ext}}, T) = c_{\mathrm{N}}|\boldsymbol{\mu}|\mathcal{L}\left(\frac{\mu_0|\boldsymbol{\mu}|(H_{\mathrm{ext}} - (\mathcal{N} - \frac{1}{3})M)}{k_{\mathrm{B}}T}\right) \quad . \tag{7.8}$$

This equation predicts that the demagnetization effects and the dipole-dipole interactions compensate each other if the sample is of spherical shape. Furthermore, it predicts that a sample should be as elongated ($\mathcal{N} \approx 0$) as possible to study differences between the magnetization behavior of different concentrated ferrofluid samples, since here the differences should be maximal.

To estimate the influence of demagnetization effects and the dipole-dipole interactions on the MPS signal, the magnetization response of samples of different shapes and filled with different concentrated ferrofluids has been simulated. The first sample is assumed to be of spherical shape ($\mathcal{N} = 1/3$) and the other one long and narrow ($\mathcal{N} \approx 0$). The iron concentration c_{Fe} of the ferrofluid has been chosen to be in the range of $0.01\,\text{mol}/1\ldots0.8\,\text{mol}/1$, which is a typical range of the MPI contrast agents. The iron concentration c_{Fe} is switched to the particle concentration c_{N} according to (7.1), since the particle concentration c_{N} and not the iron concentration c_{Fe} is needed for simulations.

The magnetic field is characterized by the drive field amplitude of $H_{\text{D}} = 0.01\text{mT}/\mu_0$ and an additional offset field, which has been linearly increased from $H_{\text{O}}^{\min} = 0\,\text{mT}/\mu_0$ to $H_{\text{O}}^{\max} = 15\,\text{mT}/\mu_0$ in $N_{\text{H}} = 50$ steps. The frequency of the magnetic field has been taken to be $\omega_0 = 2\pi \cdot 25\,\text{kHz}$ and therefore the magnetic field is given by

$$H_{\text{ext}}^k(t) = H_{\text{D}} \cdot \sin(2\pi f t) + k \cdot \frac{H_{\text{O}}^{\max}}{N_{\text{H}} - 1} \quad \text{with} \quad k = 0\ldots N_{\text{H}} - 1 \quad . \tag{7.9}$$

The particle distribution $P(d_c)$ (7.3) has been assumed to be log-normal with an expectation value of $13\,\text{nm}$ and a standard deviation of $3\,\text{nm}$, which are typical values [7], and the particle diameter has been discretized like

$$d_c(l) = 5\,\text{nm} + l \cdot \frac{28}{29} \quad \text{with} \quad l = 0\ldots29 \quad . \tag{7.10}$$

Additional simulation parameter, which have been used: Temperature $T = 300\,\text{K}$, sample volume $V_s = 10\,\mu\text{l}$, saturation magnetization of the particles $M_s = 0.6\,\text{mT}/\mu_0$.

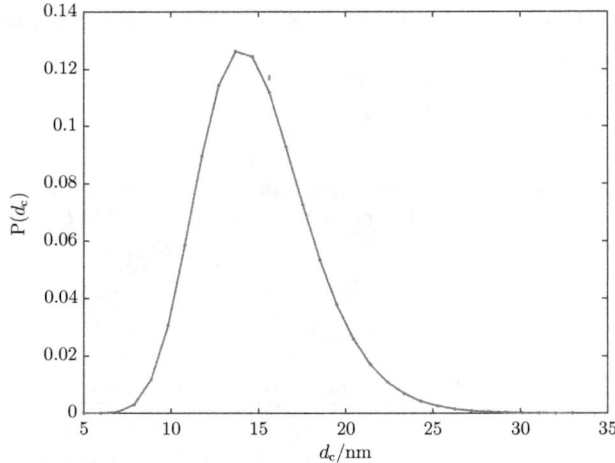

Fig. 7.3: Used particle distribution to simulate the MPS data.

The magnetization response has been simulated according to the equation

$$M(t,k) = c_{\text{N}} \sum_l |\boldsymbol{\mu}(l)| \mathcal{L}\left(\frac{\mu_0 |\boldsymbol{\mu}(l)| (H_{\text{ext}}^k(t) - (\mathcal{N} - \frac{1}{3})M(t,k))}{k_{\text{B}}T}\right) \quad , \tag{7.11}$$

where the magnetic moment has been calculated by

$$|\boldsymbol{\mu}(l)| = \frac{\pi}{6} M_s d_c(l)^3 \quad . \tag{7.12}$$

This implicit function has been solved by applying the MATLAB function `fzero`.

The Fourier coefficients

$$S_n(k) = \frac{\omega_0}{2\pi} \int_0^{2\pi/\omega_0} M(t,k) e^{-in\omega_0 t} \mathrm{d}t \tag{7.13}$$

have then been calculated with the MATLAB function `fft`. An additional sensitivity term or time derivative are not taken into consideration, since every MPS signal can be corrected by a calibration measurement.

The system function $S_n(k)$ is stored in a two-dimensional matrix, whereas the index n denotes the nth harmonic. Furthermore, the normalized system function has been calculated, which is defined by

$$|S_n(k)|_{\mathrm{norm}} = \frac{|S_n(k)| - \min_k(|S_n(k)|)}{\max_k(|S_n(k)| - \min_k(|S_n(k)|))} \quad . \tag{7.14}$$

$S_n(k)$ always ranges between 0 and 1. The advantage of normalized system function is that it allows a concentration-independent investigation of the shape of the system function.

Figure (7.5) visualizes the absolute value of the third harmonic simulated with no offset field in dependence of the iron concentration. It turns out that the relation between magnetization response of the elongated sample and the iron concentration is no longer linear for iron concentrations greater than 0.1 mol/l.

Figure (7.4) shows the evolution of the shape of the system function of fifth harmonic. The effect of the particle interactions leads to a shift of the zeros of the system function and to a stretching of their shape. For small iron concentrations the system functions of the elongated and spherical sample are indistinguishable.

One can therefore conclude, that particle interactions and demagnetization effects can be neglected if the iron concentration of the sample is below 0.1 mol/l. However, this is only a very rough estimation, since it is based on the Langevin function and therefore totally excludes any dynamical effects. Also, it excludes hydrodynamic interaction, which may be of importance [5].

Furthermore, detection of particle-particle interaction by comparing the shape of the system function measured at different concentrations appeared promising. The investigation of the deviations between the measurement data and an assumed linear dependency between the magnetization and the concentration is error-prone, since one has to prepare an exact dilution series. A small error of the dilution series would decrease the traceability of particle-particle interaction dramatically. It is therefore more promising to measure a highly diluted sample and compare its normalized system

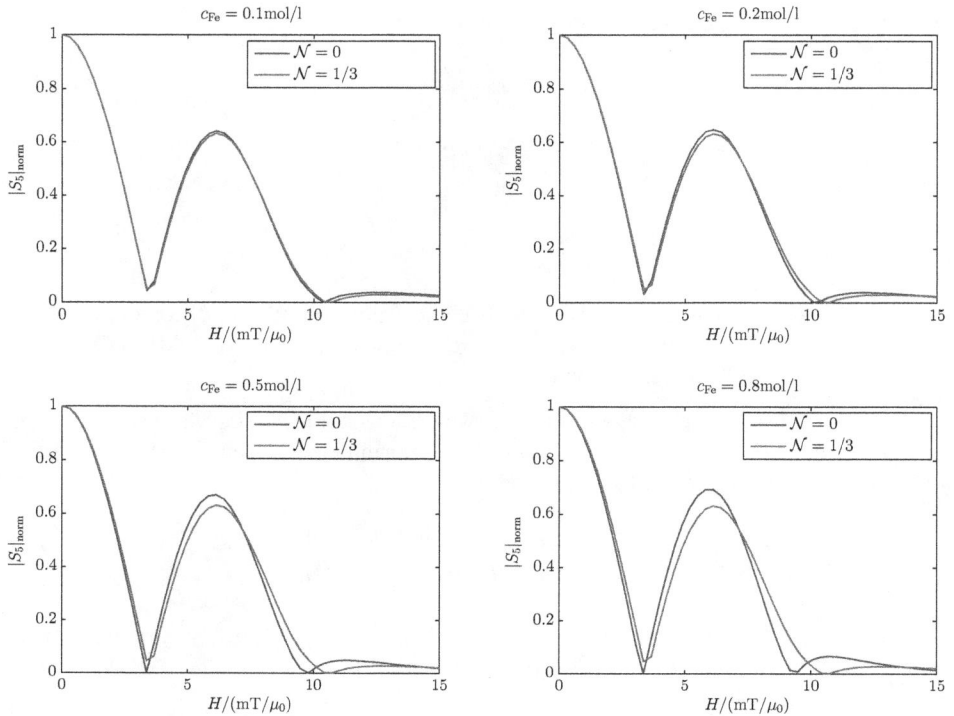

Fig. 7.4: Normalized system function of the fifth harmonic. Particle-particle interactions lead to a shift of the zeros of the system functions and a stretching of its shape. This can be used to detect particle-particle interactions.

function to the normalized system functions of any higher concentrated samples. Any deviations will then indicate particle-particle interactions. Some measurement results are presented in (8.3).

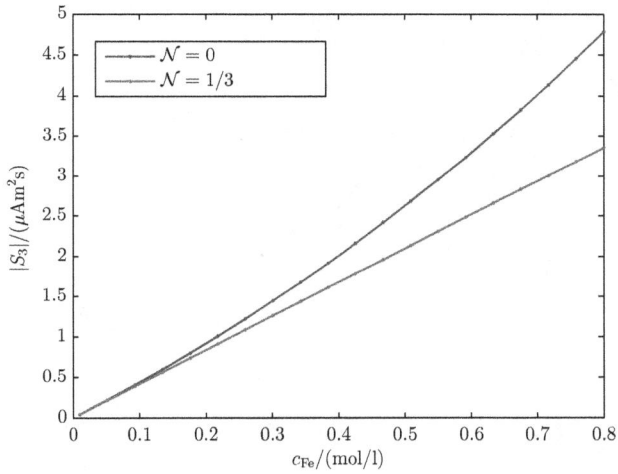

Fig. 7.5: Absolute value of the third harmonic in dependency of the iron concentration. Particle-particle interactions enhance the alignment of the magnetic moments and therefore the relation between the magnetization and the iron concentration is no longer linear above $c_{Fe} \approx 0.1 \, \mathrm{mol/l}$.

8 Measurements

8.1 MPS Experiment

The measurements have been carried out with an MPS that has been developed by the "Institut für Medizintechnik" of Lübeck (Institute of Medical Engineering). The MPS consists of two equivalent coil systems, each of them coincides with that one presented in figure (2.4). To generate an almost homogenous magnetic field, two circular transmit

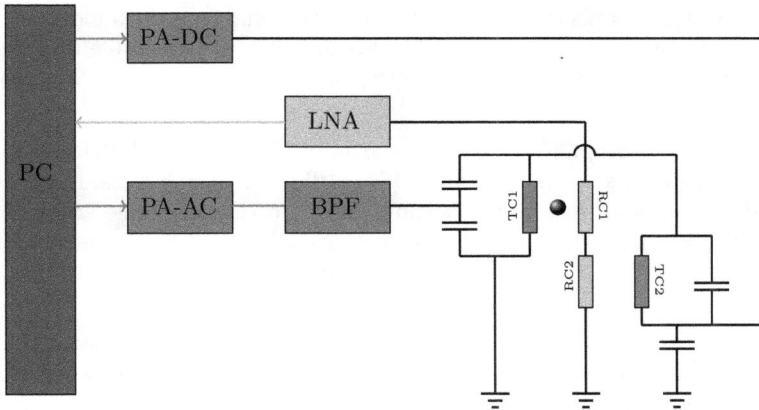

Fig. 8.1: Signal chain of the MPS. The transmitter circuit (blue) starts with a PC, which provides a sine wave. This output signal is amplified by an AC power-amplifier (PA-AC) and passes a band-pass filter (BPF) before it is supplied to the transmit coils (TC1 and TC2). In addition, TC1 and TC2 can be driven by a constant current, which is amplified by a DC power-amplifier (PA-DC). The received signal (orange) induced in the receive coil RC2 is subtracted from the signal that is induced in RC1. A ferrofluid sample is only placed in RC1 and therefore, both transmit signals cancel out. The remaining particle signal is amplified by a low noise amplifier (LNA) before it is digitized by the PC.

coils are mounted in Helmholtz configuration with a distance of 9 mm. Their outer diameter is 58 mm, their inner diameter is 19 mm and their height is 6 mm. The signal is received by a solenoid coil; its self-resonance is above 2.5 MHz and its dimensions are specified by the inner diameter of 10 mm and its length of 7 mm.

The signal diagram of the MPS is presented in figure (8.1). The transmit sine wave is generated by a signal generator in a PC and amplified by an AC power amplifier (PA-AC). The amplified signal passes a band-pass filter (BPF) before it is applied to two resonance circuits, which consist of capacitors and the two transmit coil pairs (TC1 and TC2). The generated field strengths oscillate with a frequency of 25 kHz and a maximum field strength of $40\,mT/mu_0$. Furthermore, it is possible to apply an additional offset field by the transmit coil pairs, which is also controlled by the PC and amplified by an additional power-amplifier (PA-DC). The range of the offset field is from $0\,mT/\mu_0$ to $20\,mT/\mu_0$.

Contrary to a standard MPI device (2.2), the induced signal is not processed by a band-stop filter to extract the base frequency. This is not necessary, due to the symmetric set-up of the MPS. The transmit coil pairs TC1 and the receive coil RC1 are spatially separated from the coil pairs TC2 and the receive coil RC2. The induced signal, measured by RC1, consists of the particle signal and of the directly coupled transmit signal, whereas the signal of RC2 is solely induced by the transmit signal, since a ferrofluid sample is only placed in RC1. Both receive coils are connected in series such that the signals are summed with different signs and thus both induced transmit signals get eliminated. The benefit of this design is that it is possible to measure the base frequency of the particle signal, which no longer needs to be suppressed by a BSF.

The receive signal is amplified by a low noise amplifier (LNA) and digitized by a data acquisition card with a sample rate of 5 MS/s. 125 000 transmit periods are recorded, which results in a total measurement time of 5 s. Furthermore, the acquired data is averaged over 10 transmit periods and Fourier transformed. The measured spectrum $S_{\mathrm{meas}}(f)$ is divided by a transfer function $T(f)$ of the receive channel

$$S_{\mathrm{cor}}(f) = S_{\mathrm{meas}}(f)/T(f) \tag{8.1}$$

to obtain the corrected spectrum $S_{\mathrm{cor}}(f)$, whereas the transfer function has been measured in advance. This measurement is carried out with a small test coil with a known magnetic moment, which is placed in the MPS instead of a ferrofluid sample. The transfer function is obtained by exciting the test coil and recording the induced signal at harmonics of the base frequency.

Typically, a spherical sample with a volume of $10\,\mu l$ is used, which has to be placed exactly in the center of the receive coil, since here the field is almost homogenous. This MPS is theoretically able to detect the signal of 100 harmonics of the base frequency, which is limited by the self-resonance frequency of the receive coil and the sampling rate of the data acquisition card.

8.2 Contrast Agent

Resovist® has been used as contrast agent, which is produced by the Bayer Schering Pharma AG, Germany. It is a ferrofluid of superparamagnetic iron oxide particles

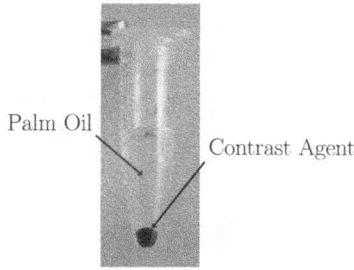

Fig. 8.2: A typical sample consist of an Eppendorf Safe-Lock tube, filled with palm oil and the contrast agent.

(SPIO), which are coated with a biocompatible shell. According to the manufacturer's data *Resovist*® has an iron concentration of $c_{Fe} = 0.5$mol/l. This contrast agent is clinically proven, because it has been originally designed as a liver specific magnetic resonance contrast agent [60].

Fortunately, Resovist also shows a good performance with respect to MPI and it is therefore often used [32, 7].

A sample is prepared by filling $V_s = 10\,\mu l$ of *Resovist*® into an Eppendorf Safe-Lock tube. In addition, $100\,\mu l$ of palm oil are filled on the top of the ferrofluid. This prevents the ferrofluid from drying out and furthermore it fixates the ferrofluid, because the water and the oil phase do not mix. All measurements of this contrast agent have been carried out at room temperature.

8.3 Measurement Results

8.3.1 Measurement of a Dilution Series

The aim of this measurement has been to investigate the relationship between the measured magnetization and the concentration of the ferrofluid sample. Different concentrations of the ferrofluid samples have been obtained by diluting *Resovist*® with water. Eight samples have been prepared as specified by the following dilution factors

$a_1 = 1$, $a_2 = 0.875$, $a_3 = 0.75$, $a_4 = 0.625$, $a_5 = 0.5$, $a_6 = 0.375$, $a_7 = 0.25$, $a_8 = 0.125$.

The dilution factor is defined by the ratio

$$a = \frac{V_{Res}}{V_s} \tag{8.2}$$

between the volume of Resovist in the sample V_{Res} and the volume of the sample V_s. First, standard samples have been measured, which are approximately of spherical

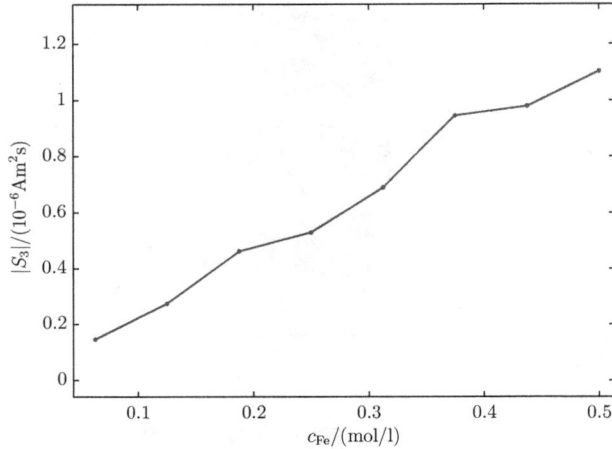

Fig. 8.3: Absolute value of the third harmonic in dependence of the iron concentration of the Resovist samples.

shape (8.2). The measurement has been carried out under a drive field strength of $H_D = 10\,\text{mT}/\mu_0$ and additional offset fields between $0\,\text{mT}/\mu_0$ and $0.01\,\text{mT}/\mu_0$. As presented in figure (8.3), the magnetization increases linearly with the concentration of the sample, whereas the small deviation is presumably caused by imperfections during sample preparations.

In addition, the normalized system functions have been calculated. The absolute values of the 5th and the 6th harmonic are presented in (8.4). It is apparent that the structure of the system function is indeed dominated by Chebyshev polynomials of the second kind (2.23). Furthermore, there is only a very small difference between the shape of the system function of the low and the high concentrated sample.

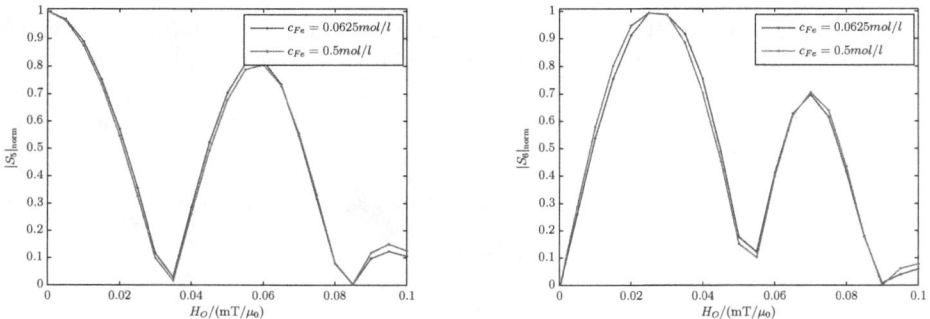

Fig. 8.4: Absolute values of the 5th and the 6th harmonic of two different concentrated samples, measured with different offset fields.

Both results coincide well with the Weiss model (7), which predicts that, in the case of a spherical sample, demagnetization effects and particle-particle interactions compensate each other.

An additional attempt has been made to carry out a more meaningful measurement by measuring an elongated sample. Here, tips of pipettes have been used as ferrofluid container. These tips are of cylindrical shape, 7 mm in length and the inner diameter is about 1.5 mm. However, strange effects occurred, while measuring these samples, which did not depend on the concentration of the ferrofluid. Furthermore, the reproducibility of the measurements was no longer assured, although the appearance of the sample did not change during two different measurement series. It may be possible that these samples are too small and narrow, so that surface or agglomeration effects are important.

8.3.2 Estimated Particle Distribution

The genetic fitting algorithm has been applied as described in (5.4), whereby only the first 15 harmonics have been considered. The measurement data has been acquired by a measurement carried out under a drive field amplitude of $H_D = 0.01 \, \text{mT}/\mu_0$ and no offset fields. In contrast, the steady state solutions of (5.17) have been calculated with the same drive field amplitude and a small additional offset field. This is necessary, because of the terrestrial magnetic field. The value of this offset field is $H_O = 48.5 \, \mu\text{T}/\mu_0$ and has been found by simply trying different offset field values and comparing the results of the fitting algorithm. The strength of this magnetic field is in good agreement with typical values of the terrestrial magnetic field in Germany.

The King solution, presented in figure (8.5), has been found after the genetic algorithm has completed 1500 generation cycles. Even though the fitness has been judged by comparing the amplitude spectrum and the imaginary part of the spectrum of the simulated data against the measurement data, it is of course also possible to compare the time signals and phase spectra of the simulated and the measurement data, which is shown in figure (8.6). The estimated particle distribution is presented in figure (8.7), whereas the figure (8.8) visualizes the marginal distributions of d_c and τ_0.

According to (5.27), the number of particles has been estimated to be $N \approx 10^{13.1}$ and therefore the iron concentration of the sample has been estimated to be (7.1)

$$c_{Fe} = \frac{3\rho_{Fe_3O_4}}{M_{Fe_3O_4}} \frac{N}{V_s} \frac{\pi}{6} \sum_{i=1}^{N_d} \sum_{j=1}^{N_\tau} P(d_c(i)\tau_0(j))d_c(i)^3 \approx 0.147 \frac{\text{mol}}{\text{l}} \quad . \tag{8.3}$$

The mean core diameter and the mean relaxation time determined by the King distribution are given by

$$\begin{aligned} \langle d_c \rangle &= 14.5 \, \text{nm} \\ \langle \tau_0 \rangle &= 2.6 \, \mu\text{s} \quad . \end{aligned} \tag{8.4}$$

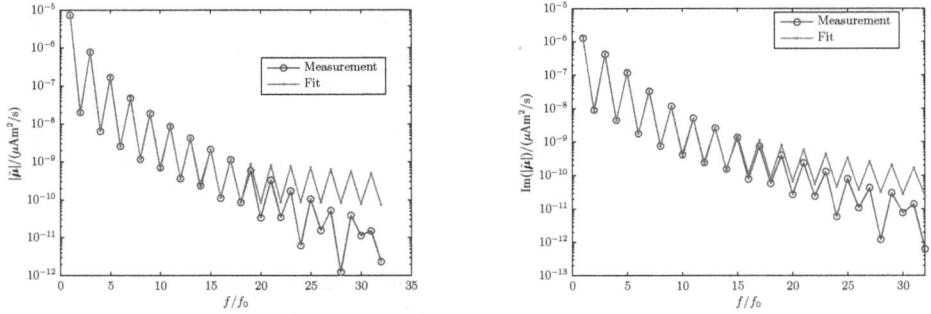

Fig. 8.5: The absolute spectrum and absolute value of the imaginary spectrum of the King solution compared to the measurement data.

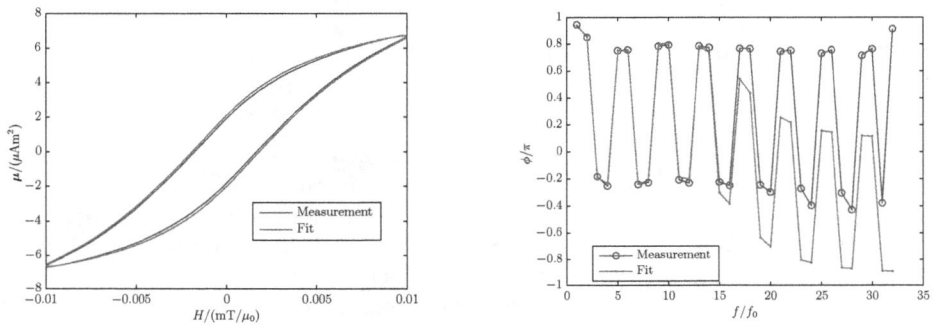

Fig. 8.6: Time signal and phase spectrum of the imaginary spectrum of the King solution compared to the measurement data.

Furthermore, it is possible to calculate the saturation magnetization of the sample (2.1) according to

$$M_{\text{sat}} = 960 \frac{\text{A}}{\text{m}} \quad . \tag{8.5}$$

The result of the estimated iron concentration differs from the manufacture's data. However, it was not expected that this approximated particle model is able to provide quantitatively correct values, due to remaining uncertainties with respect to the magnetic moment of a particle (3.2.4) and the versatile particle dynamics (6).

Although this result is encouraging, an extrapolation to measurement data, which are measured with additional offset fields, yet fails, see Fig. (8.9). Additional attempts have been made to estimate the particle distribution by including measurement data in the fitting process, which have been measured with offset fields. However, this has not been successful.

It is not yet clear, why this extrapolation fails. It may be reasoned in an insufficient particle model or in not correctly calibrated offset fields.

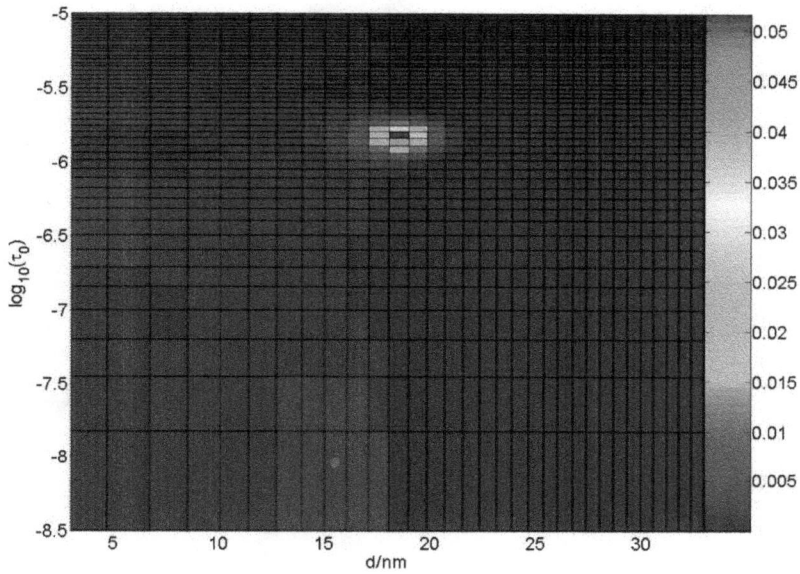

Fig. 8.7: Estimated King distribution.

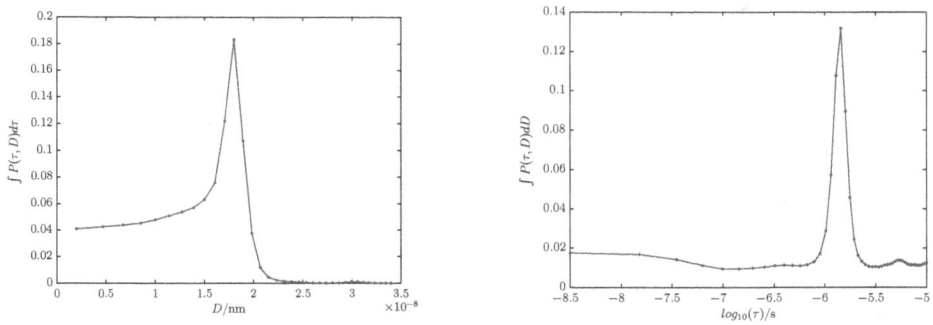

Fig. 8.8: Marginal distributions of the King distribution.

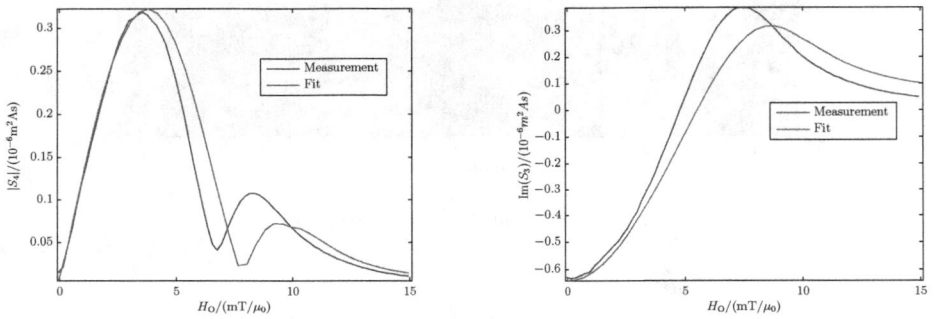

Fig. 8.9: Extrapolated King solution to measurement data, which have been measured with additional offset fields.

9 Conclusions

In this work the physical properties of the contrast agents, which are ferrofluids, with respect to magnetic particle imaging (MPI) have been studied. Since MPI is a contrast agent based imaging technology it is indispensable to understand its physical properties.

Therefore, an introduction into the physics of the MPI contrast agent is given with a detailed derivation of the Langevin function. Previous works often consider the magnetic particles to be in thermodynamic equilibrium, which is a rough approximation due to the time varying magnetic field of an MPI scanner and the finite relaxation times due to the Brownian and Néel diffusion.

To be able to simulate the particle dynamics of the MPI contrast agents a comprehensive derivation of a stochastic particle model is presented. This particle model is based on Langevin equations and incorporates both, the Brownian and the Néel diffusion. First the Brownian and Néel diffusion have been studied separately. For each diffusion process the Fokker Planck equations have been derived. The equilibrium solutions of the Fokker Planck equations have then be connected to the Boltzmann statistics to determine the diffusion constants. In the case of the Brownian diffusion the Fokker Planck equation has been solved, whereas it turns out that the approach of solving the Fokker Planck equation is not practicable with respect to MPI simulation studies. Therefore, the Euler-Maruyama and the Heun integration scheme have been implemented to numerically integrate the Langevin equations of the Brownian and the Néel diffusion. Both integration schemes have been verified by comparing their simulation results of equilibrium particle ensembles to the solution of the distributions, which are predicted by the Boltzmann statistic. Furthermore, a detailed analysis of the time-step dependency of each integration scheme has been performed, whereas it turns out, that the Brownian diffusion process should be integrated by the Euler-Maruyama solver and the Néel diffusion process by the Heun solver. Both diffusion processes are then combined by introducing a new integration scheme. The integration scheme has been verified by again comparing its equilibrium solutions to the Boltzmann statistic. Furthermore, simulations have been performed to validate the limits of the rigid and the blocked particles.

Whereas this stochastic particle model seems to be capable to give a deep insight into the dynamics of the contrast agent it is not appropriate to fit measurement data. Therefore, an approximated particle model (ODE-based particle model), based on ordinary differential equations, is introduced and adjusted by comparing its solution to the solution of the Brownian diffusion. In addition, an extended smooth genetic fitting algorithm is proposed to fit this approximated particle model to measurement data.

The new fitting algorithm and the ODE-based particle model have then been applied to approximate measurement data, carried out by a magnetic particle spectrometer (MPS). It turns out that promising results are obtained without offset fields. The modeling of measurement data including additional offset fields unfortunately fails. However, reasonable extensions are proposed, so that the combination of the fitting algorithm and the ODE-based particle model may be helpful to generate the system function of any arbitrary MPI scanner.

The implemented stochastic particle model and the ODE-based particle model have been applied to investigate the question, whether there is an optimal base frequency of an MPI scanner. It has been found that the particle signal depends linearly on the frequency, whereas the particle anisotropy has a strong impact on the signal intensity.

Beside the simulation-based acquisition of the MPI system function there is the measurement-based approach. Here the signal to noise ratio depends linearly on the concentration of the contrast agent. Therefore a simulation study has been performed to estimate the specific concentration of a ferrofluid up to which it is possible to safely neglect particle-particle interactions. In addition, a more sensitive analysis method is proposed to detect particle-particle interactions.

Further extensions are needed with respect to the ODE-based particle model. However, the general idea of the ODE-based particle model in combination with the proposed fitting algorithm and the detailed derivation of the stochastic particle model are hopefully useful for future developments in MPI technology.

10 Bibliography

[1] A. Aharoni. *Introduction to the theory of ferromagnetism.* Oxford University Press, 2 edition, 2000.

[2] J.Weizenecker B. Gleich. Tomographic imaging using the nonlinear response of magnetic particles. *Nature*, 453:1214–1217, 2005.

[3] I. A. Bakelaar B. W. M. Kuipers. Complex magnetic susceptibility setup for spectroscopy in the extremely lowfrequency range. *Review of Scientifig Instruments*, 79, 2007.

[4] C. D. Graham B.D. Cullity. *Introduction to magnetic materials.* Wiley-VCH, 2009.

[5] D V Berkov, N L Gorn, R Schmitz, and D Stock. Langevin dynamic simulations of fast remagnetization processes in ferrofluids with internal magnetic degrees of freedom. *Journal of Physics: Condensed Matter*, 18(38):S2595, 2006.

[6] Gorn Berkov. *Handbook of Magnetism and Advanced Magnetic Materials*, volume 2, chapter Numerical Simulation of Quasistatic and Dynamic Remagnetization Processes with Special Applications to Thin Films and Nanoparticles. Springer, 2006.

[7] S Biederer, T Knopp, T F Sattel, K Lüdtke-Buzug, B Gleich, J Weizenecker, J Borgert, and T M Buzug. Magnetization response spectroscopy of superparamagnetic nanoparticles for magnetic particle imaging. *Journal of Physics D: Applied Physics*, 42(20):205007, 2009.

[8] J. Bohnert, B. Gleich, J. Weizenecker, J. Borgert, and O. Dössel. Evaluation of induced current densities and sar in the human body by strong magnetic fields around 100 khz. In *4th European Conference of the International Federation for Medical and Biological Engineering*, volume 22 of *IFMBE Proceedings*. Springer Berlin Heidelberg, 2009.

[9] H.B. Braun. *Structures and Dynamics in Heterogenous Systems*, chapter Stochastic Magnetization Dynamics in Magnetic Nanostructures: From Néel-Brown to Soliton-Antisoliton Creation. World Scientific, 2000.

[10] Soshin Chikazumi and Jr. *Physics of Ferromagnetism (International Series of Monographs on Physics)*. Oxford University Press, 1997.

[11] R. M. Cornell and U. Schwertmann. *The Iron oxides : structure, properties, reactions, occurences and uses (2nd ed.)*. Wiley-VCH, 2nd edition, 2003.

[12] J. Villain D. Gatteshi, R. Sessoli. *Molecular Nanomagnetis*. Oxford University Press, 2006.

[13] W. Demtröder. *Experimentalphysik 2*. Springer, 3 edition, 2004.

[14] J.K.G. Dhont. *An Introduction to Dynamics of Colloids*. ELSEVIER, 1996.

[15] Patrick S. Doyle and Patrick T. Underhill. Brownian dynamics simulations of polymers and soft matter. In Sidney Yip, editor, *Handbook of Materials Modeling*, pages 2619–2630. Springer Netherlands, 2005.

[16] M. Zakai E. Wong. On the convergence of ordinary integrals to stochastic integrals. *The Annals of Mathematical Statistics*, 36:1560–1564, 1965.

[17] P. C. Fannin. Magnetic spectroscopy as an aide in understanding magnetic fluids. *Journal of Magnetism and Magnetic Materials*, 252:59 – 64, 2002.

[18] P. C. Fannin. Characterisation of magnetic fluids. *Journal of Alloys and Compounds*, 369(1-2):43 – 51, 2004. Proceedings of the VI Latin American Workshop on Magnetism, Magnetic Materials and their Applications.

[19] Walter S. D. Folly and Ronaldo S. de Biasi. Determination of particle size distribution by FMR measurements. *Brazilian Journal of Physics*, 31:398 – 401, 09 2001.

[20] José Luis García-Palacios and Francisco J. Lázaro. Langevin-dynamics study of the dynamical properties of small magnetic particles. *Phys. Rev. B*, 58(22):14937–14958, Dec 1998.

[21] C.W. Gardiner. *Handbook of Stochastic Methods*. Springer, 3rd edition, 2004.

[22] B Gleich, J Weizenecker, and J Borgert. Experimental results on fast 2d-encoded magnetic particle imaging. *Physics in Medicine and Biology*, 53(6):N81, 2008.

[23] P.W. Goodwill and S.M. Conolly. The x-space formulation of the magnetic particle imaging process: 1-d signal, resolution, bandwidth, snr, sar, and magnetostimulation. *IEEE Trans Med Imaging*, 29(11):1851–9, 2010.

[24] S. P. Gubin, editor. *Magnetic Nanoparticles*. Wiley-VCH, 2009.

[25] A.P. Guimaraes. *Principles of Nanomagnetism*. Nanoscience and Technology. Springer, 2009.

[26] M W Gutowski. Smooth genetic algorithm. *Journal of Physics A: Mathematical and General*, 27(23):7893, 1994.

[27] J. Borgert I. Schmale, B. Gleich. Jfet noise modelling for mpi receivers. In *Magnetic Nanoparticles*. World Scientific, 2010.

[28] P.W. Selwood I. Yasumuri, D. Reinen. Anisotropic behaviour in superparamagnetic systems. *Journal of Applied Physics 12*, 34, 1963.

[29] A. O. Ivanov. Manetic properties of plolydisperse ferrofluids: A critical comparison between experiment, theory and computer simulation. *Physical Review E*, 75, 2007.

[30] B. Gleich J. Borgert J. Rahmer, J. Weizenecker. Signal encoding in magnetic particle imaging: properties of the system function. *BMC Medical Imaging*, 2009.

[31] B. Gleich J. Weizenecker, J. Bogert. A simulation study on the resolution and sensitivity of magnetic particle imaging. *Physics in Medicine and Biology*, 52:6363–6374, 2007.

[32] J. Rahmer H. Dahnke J. Bogert J. Weizenecker, B. Gleich. Three-dimensional real-time in vivo magnetic particle imaging. *Physics in Medicine and Biology*, 54, 2009.

[33] J. Rahmer J. Bogert J. Weizenecker, B. Gleich. Particle dynamics of mono-domain paritcles in magnetic particle imaging. In *Magnetic Nanoparticles*. World Scientific, 2010.

[34] NG Van Kampen. *Stochastic processes in physics and chemistry*. North Holland, 2007.

[35] L B Kiss, J Söderlund, G A Niklasson, and C G Granqvist. New approach to the origin of lognormal size distributions of nanoparticles. *Nanotechnology*, 10(1):25, 1999.

[36] T. Knopp. *Effiziente Rekonstruktion und alternative Spulentopologien für Magnetic-Particle-Imaging*. PhD thesis, Universität zu Lübeck, 2010.

[37] T. Knopp, S. Biederer, T. Sattel, J. Weizenecker, B. Gleich, J. Borgert, and T.M. Buzug. Trajectory analysis for magnetic particle imaging. *Physics in Medicine and Biology*, 54(2):385–397, 2009.

[38] T. Knopp, T. Sattel, S. Biederer, J. Rahmer, J. Weizenecker, B. Gleich, J. Borgert, and T.M. Buzug. Model-based reconstruction for magnetic particle imaging. *IEEE Trans. Med. Imag.*, 29(1):12–18, 2010.

[39] T Knopp, T F Sattel, S Biederer, and T M Buzug. Field-free line formation in a magnetic field. *Journal of Physics A: Mathematical and Theoretical*, 43(1):012002, 2010.

[40] T. Knopp, T. F. Sattel, S. Biederer, and T. M. Buzug. Limitations of measurement-based system functions in magnetic particle imaging. In *SPIE Medical Imaging*, volume 76261F, 2010.

[41] Helmut Kronmueller. *General Micromagnetic Theory*. John Wiley & Sons, Ltd, 2007.

[42] Ryogo Kubo and Natsuki Hashitsume. Brownian motion of spins. *Progress of Theoretical Physics Supplement*, 46:210–220, 1970.

[43] A. Lange. Thermomagnetic convection of magnetic fluids in a cylindrical geometry. *Physics of fluids*, 14(7), 2002.

[44] Nicola Bruti Liberati and Eckhard Platen. On the efficiency of simplified weak taylor schemes for monte carlo simulation in finance. Research Paper Series 114, Quantitative Finance Research Centre, University of Technology, Sydney, 2004.

[45] W. Lowrie. *Fundamentals of geophysics*. Cambridge University Press, 2 edition, 2007.

[46] J. Davis M. Galassi. *GNU Scientific Library Reference Manuel*. Network Therory Limited, 2009.

[47] George Marsaglia. http://www.stat.fsu.edu/pub/diehard/.

[48] V.I. Stepanov M.I. Shliomis. *Relaxation Phenomena in Condensed Matter*, chapter Theory of the Dynamic Susceptibility of Magnetic Fluids. John Wiley & Sons, 1994.

[49] Romisa Asadi Shareen H. Doak Neenu Singh, Gareth J.S. Jenkins. Potential toxicity of superparamagnetic iron oxide nanoparticles (spion). *Nano Reviews*, 2010.

[50] U. Nowak and D. Hinzke. Magnetic nanoparticles: The simulation of thermodynamic properties. In Bernhard Kramer, editor, *Advances in Solid State Physics*, volume 41 of *Advances in Solid State Physics*, pages 613–622. Springer, 2001.

[51] E. Platen P. E. Kloeden. *Numerical Solution of Stochastic Differential Equation*. Springer, 2 edition, 1995.

[52] P. P. Stang P. W. Goodwill, G. C. Scott and S. M. Conolly. Narrowband magnetic particle imaging. *Medical Imaging*, 28, 2009.

[53] S.S. Papell. Low viscosity magnetic fluid obtained by the colloidal suspension of magnetic particles, 1963.

[54] S. E. Haupt R. L. Haupt. *Practical Genetic Algorithms*. John Wiley & Sons, 2004.

[55] A. P. Khandhar K. M. Krishnan R. M. Ferguson, K. R. Minard. Optimizing magnetite nanoparticles for mass sensitivity in magnetic particle imaging. *Medical Physics*, 2011.

[56] K. M. Krishnan R. M. Ferguson, K. R. Minard. Optimization of nanoparticle core size for magnetic particle imaging. *Journal of Magnetism and Magnetic Materials*, 2009.

[57] Adam M Rauwerdink, Eric W Hansen, and John B Weaver. Nanoparticle temperature estimation in combined ac and dc magnetic fields. *Physics in Medicine and Biology*, 54(19):L51, 2009.

[58] Adam M. Rauwerdink and John B. Weaver. Measurement of molecular binding using the brownian motion of magnetic nanoparticle probes. *Applied Physics Letters*, 96(3):033702, 2010.

[59] Adam M. Rauwerdink and John B. Weaver. Viscous effects on nanoparticle magnetization harmonics. *Journal of Magnetism and Magnetic Materials*, 322(6):609 – 613, 2010.

[60] Peter Reimer and Thomas Balzer. Ferucarbotran (resovist): a new clinically approved res-specific contrast agent for contrast-enhanced mri of the liver: properties, clinical development, and applications. *European Radiology*, 13, 2003.

[61] W. Rüemelin. Numerical treatment of stochastic differential equations. *SIAM Journal on Numerical Analysis*, 19, 1982.

[62] H. Risken. *The Fokker-Planck Equation*. Springer, 2nd edition, 1989.

[63] R.E. Rosensweig. *Ferrohydrodynamic*. Cambridge University Press, 1985.

[64] S.Asmussen. *Stochastic Simulation Algorithms and Analysis, Springer (2007)*. Springer, 2007.

[65] C. Scherer. Stochastic molecular dynamics of colloidal particles. *Brazilian Jounal of Physics*, 34(2A), 2004.

[66] I. Schmale, B. Gleich, J. Kanzenbach, J. Rahmer, J. Schmidt, J. Weizenecker, and J. Borgert. An introduction to the hardware of magnetic particle imaging. In Ratko Magjarevic, Olaf Dässel, and Wolfgang C. Schlegel, editors, *World Congress on Medical Physics and Biomedical Engineering, September 7 - 12, 2009, Munich, Germany*, volume 25/2 of *IFMBE Proceedings*, pages 450–453. Springer Berlin Heidelberg, 2009.

[67] T. Schrefl, J. Fidler, D. Suess, W. Scholz, and V. Tsiantos. Micromagnetic simulation of dynamic and thermal effects. In Yi Liu, David J. Sellmyer, and Daisuke Shindo, editors, *Handbook of Advanced Magnetic Materials*, pages 128–146. Springer US, 2006.

[68] M I Shliomis. Magnetic fluids. *Soviet Physics Uspekhi*, 17(2):153, 1974.

[69] R. Skomski. *Simple Models of Magnetism*. Oxford University Press, 2008.

[70] R. Skomski and D. J. Sellmyer. Intrinsic and extrinsic properties of advanced magnetic materials. *ChemInform*, 37(47), 2006.

[71] L. Trahms. *Colloidal Magnetic Fluids*, chapter Biomedical Applications of Magnetic Nanoparticles, pages 327–358. Springer-Verlag Berlin Heidelberg, 2009.

[72] K. D. Usadel. Temperature-dependent dynamical behavior of nanoparticles as probed by ferromagnetic resonance using landau-lifshitz-gilbert dynamics in a classical spin model. *Phys. Rev. B*, 73(21):212405, Jun 2006.

[73] J Weizenecker, J Borgert, and B Gleich. A simulation study on the resolution and sensitivity of magnetic particle imaging. *Physics in Medicine and Biology*, 52(21):6363, 2007.

[74] Juergen Weizenecker, Bernhard Gleich, and Joern Borgert. Magnetic particle imaging using a field free line. *Journal of Physics D: Applied Physics*, 41(10):105009, 2008.

[75] J.T. Waldron W.T. Coffey, Yu. P.Kalmykov. *The Langevin Equation*. World Scientific, 2 edition, 2004.

[76] Takashi Yoshida and Keiji Enpuku. Simulation and quantitative clarification of ac susceptibility of magnetic fluid in nonlinear brownian relaxation region. *Japanese Journal of Applied Physics*, 48(12):127002, 2009.

Infinite Science Publishing provides a publication platform for excellent theses as well as scientific monographies and conference proceedings for reasonable costs.

These publications enable scientists and research organizations to reach the maximum attention for their results.

The service of Infinite Science Publishing comprises the entire range from the publication of print-ready documents up to cover design as well as copy-editing of single articles.

Infinite Science Publishing is an imprint of the Infinite Science GmbH, a University of Lübeck spin-off and service partner of the BioMedTec Science Campus.

www.infinite-science.de/publishing

Infinite Science GmbH
MFC 1 | BioMedTec Wissenschaftscampus
Maria-Goeppert-Str. 1, 23562 Lübeck
book@infinite-science.de

Infinite Science
Publishing

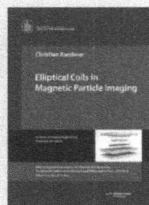

www.ingramcontent.com/pod-product-compliance
Lightning Source LLC
Chambersburg PA
CBHW081109220326
41598CB00038B/7290